Principles of Automatic Transmissions

Jay Webster

Associate Professor

Department of Industrial Education

California State College at Long Beach

Prakken Publications, Inc.

Ann Arbor, Michigan

Printed in the United States of America

Library of Congress Catalog Card Number 74-83263

ISBN: 0-911168-32-X

Second Printing 1975

Prakken Publications, Inc.
416 Longshore Drive, P.O. Box 623, Ann Arbor, Michigan 48107
Publishers of *School Shop* magazine and *The Education Digest*

To the reader:

This book is organized differently from many you have read. Each Unit is divided into two parts: text and workbook. The text sections are organized to explain the basic principles of automatic transmission operation. A workbook section follows each text section. The workbook materials are designed to help you learn the important points covered in the text. The workbook materials are self-instructional. That is, you can teach yourself if you use the materials correctly. The workbook is made up of a variety of questions and problems. Try to answer each of the questions. Then check yourself by turning to the Answers to Questions that follow each workbook section. You will find both the correct answer and an explanation.

Acknowledgments:

I would like to thank the following companies for their kind permission to reproduce photographs and art work: Buck Engineering Company, Inc., Chrysler Corporation, Ford Motor Company, General Motors Corporation, Hayden Trans-Cooler, Inc., *Trailer Life Magazine*, and Vickers Division of Sperry Rand Corporation.

And thanks to Roland Gehring, Petaluma High School; Dewey Stone, Santa Ana College; and Ken Welch, Esperanza High School, for their helpful comments on the manuscript.

Thanks are due also to those representatives of the automotive industry who reviewed the manuscript, and especially to Jon E. Stentstrom and Chester S. Potters.

Jay Webster

Long Beach, California, 1974

Table of Contents

Section I
INTRODUCTION TO AUTOMATIC TRANSMISSIONS

Section II
FLUID COUPLINGS AND TORQUE CONVERTERS

Section III
PLANETARY GEARBOX

Section IV
THE HYDRAULIC CONTROL SYSTEM

Section V
MAINTENANCE AND SERVICE

Section I

INTRODUCTION TO AUTOMATIC TRANSMISSIONS

What the Automatic Transmission Is and Does

Importance of Automatic Transmissions

The study of automatic transmissions is becoming more and more important. A large percentage of today's passenger vehicles are equipped with an automatic transmission. More and more heavy vehicles such as trucks and buses are being equipped with automatics.

There is, and will continue to be, a high demand for mechanics who understand and can service automatic transmissions. Even the mechanic who services other vehicle components must have a working knowledge of automatic transmission operation. The transmission affects engine performance to the extent that the tune-up technician must be able to diagnose vehicle performance problems related to the transmission. The chassis expert will need to know what noises and vibrations can be caused by the transmission. The vehicle's cooling, electric, and vacuum systems may be interconnected with the transmission. No mechanic can afford to be ignorant about automatic transmissions since sometimes a diagnosis of transmission malfunction is made when a tune-up is all that is really needed.

The Transmission's Job

The transmission's job is to help the engine get the vehicle moving from a standstill and to help it accelerate to road speed. The engine and transmission work as a team.

The rear wheels of the vehicle rotate to drive the vehicle. The twisting or turning effort to rotate the tires is called torque. The torque is developed by the engine. The problem is that the engine alone cannot develop enough torque to get the vehicle moving from a standstill. The engine also needs help during acceleration and hill climbing.

The torque that a typical V-8 engine can produce can be measured and plotted on a graph similar to the one in figure 1-1. The unit of measurement

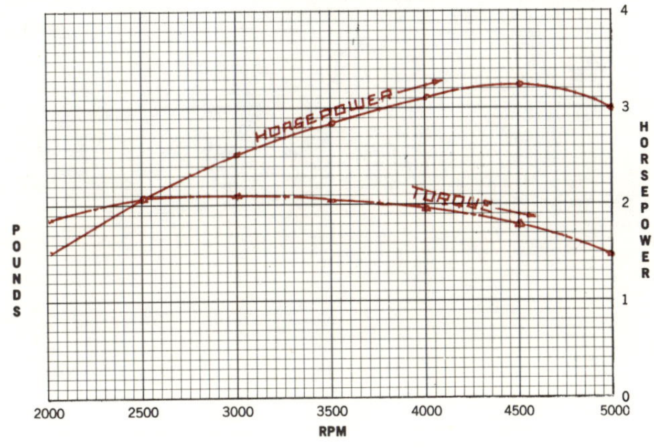

Figure 1-1

Typical Engine Torque Curve
(Buck Engineering Co., Inc.)

for torque is pounds-feet. The torque is shown at various engine speeds (rpm). Different engines, of course, describe different curves, but they are all similar in this respect—at low engine rpm torque is low, and as engine rpm increases, torque increases up to a point in the rpm range when the engine will have trouble breathing in enough air and fuel. At this point the torque curve will begin to drop off. The transmission must work to overcome this low torque at low rpm situation. The transmission at the lower gear ratios also allows the engine to operate at higher speeds during low vehicle speeds so the engine can operate at, or near, its best torque range. The transmission's job is to multiply the engine's low initial torque to get the vehicle moving and to meet varying road conditions.

The Automatic Transmission

The automatic transmission is a device which automatically adjusts drive-line torque to meet vehicle operating conditions. Most of today's automatic transmissions are basically the same. They are composed of the three main sections, shown in figure 1-2.

The Torque Converter. The torque converter serves much the same function as the foot-operated clutch on a manual transmission equipped vehicle. The torque converter allows the vehicle to idle without moving. It also helps in the job of multiplying the engine's torque.

The Planetary Gearbox. The planetary gearbox serves the same function as the gears in a standard transmission. The planetary gears provide a means of varying the gear ratio between the engine and the rear wheels in order to multiply torque to meet driving requirements.

The Hydraulic Controls. The hydraulic control system is the "brain" of the transmission. This system operates to sense and select the correct degree of torque multiplication for every driving requirement.

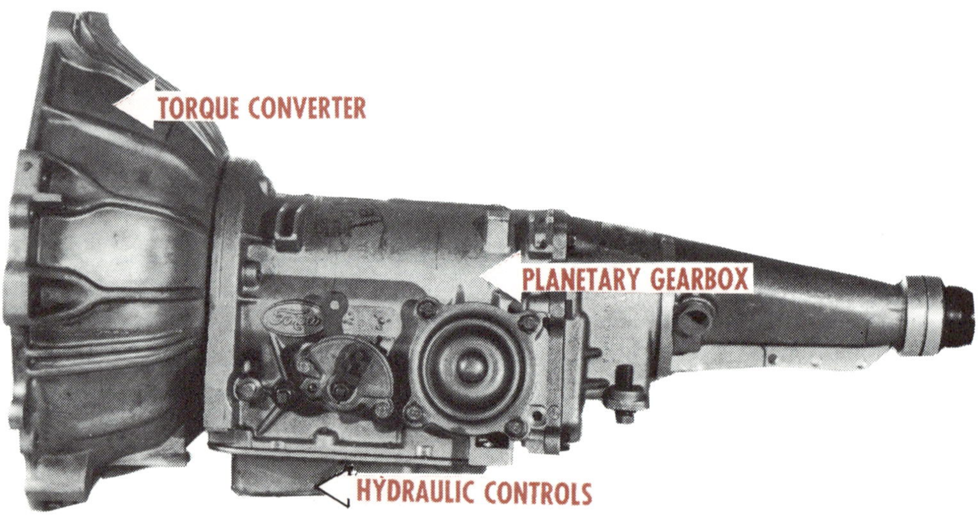

Figure 1-2

Three Main Sections of an Automatic Transmission

UNIT 1
What the Automatic Transmission Is and Does

True-False Questions (Indicate the correct choice)

1. The study of automatic transmissions is becoming more and more important. T F

2. The transmission's job is to help the engine get the vehicle moving and to help in acceleration. T F

3. Twisting or turning effort developed by the engine is called *power.* T F

4. The torque of a typical V-8 engine is high at low rpm. T F

5. The unit of measurement for torque is pounds-feet. T F

6. Engine torque increases indefinitely with rpm. T F

Parts Identification

The automatic transmission is composed of three main components. Identify the major components.

1.________________ 2.________________ 3.________________

Multiple Choice Questions (Indicate the correct choice)

1. The automatic transmission is linked to the vehicle's:
   a. cooling system
   b. electrical system
   c. vacuum system
   d. all of the above

2. At low engine rpm the torque of a typical V-8 engine would be:
   a. low
   b. medium
   c. high

3. At high engine rpm the torque curve of a typical V-8 engine is:
   a. increasing
   b. decreasing
   c. leveled off

4. Which of the following is *not* one of the three main components of an automatic transmission?
   a. torque converter
   b. hydraulic controls
   c. foot-operated clutch
   d. planetary gearbox

5. The component of an automatic transmission which serves the same function as a clutch on a manual transmission equipped vehicle is a:
   a. torque converter
   b. planetary gearbox
   c. pump
   d. band

6. The job of multiplying torque to meet driving requirements is accomplished by:
   a. torque converter
   b. planetary gearbox
   c. hydraulic control system
   d. both *a* and *b*

7. The device which operates to sense and select the correct degree of torque multiplication is:
   a. torque converter
   b. planetary gearbox
   c. hydraulic control system

UNIT 1
What the Automatic Transmission Is and Does

True-False Questions

1. This statement is true. A knowledge of automatic transmissions is becoming increasingly important.

2. This statement is true. The engine and transmission work as a team to get the vehicle moving and to help it accelerate.

3. This statement is false. Twisting or turning effort developed by the engine is called *torque* (not *power*).

4. This statement is false. A typical V-8 engine has *low* torque at low rpm (not high torque at low rpm).

5. This statement is true. Torque is expressed in pounds-feet.

6. This statement is false. Engine torque, rather than increasing indefinitely with rpm, increases until the engine's breathing becomes difficult, then it falls off.

Parts Identification

The three main components of an automatic transmission are:

1. torque converter   2. planetary gearbox   3. hydraulic controls

Multiple Choice Questions

1. The correct choice is *d.* The automatic transmission is linked to the vehicle's cooling, electrical, and vacuum systems.

2. The correct choice is *a.* At low engine rpm the torque of a typical V-8 engine would be low.

3. The correct choice is *b.* The engine has more difficulty breathing at high rpm, and the torque decreases.

4. The correct choice is *c*. The main components of an automatic transmission are the torque converter, hydraulic controls, and planetary gearbox.

5. The correct choice is *a*. The torque converter allows the vehicle to idle without moving.

6. The correct choice is *d*. Both the torque converter and the planetary gearbox operate to multiply torque to meet driving requirements.

7. The correct choice is *c*. The hydraulic control system is the "brain" of the automatic transmission.

Fluids for Automatic Transmissions

The automatic transmission is a hydraulically-operated device. This means that it must depend on fluids for its operation. The functions the fluid must perform, as well as its properties, composition, and types, are presented in this unit.

The Functions of Automatic Transmission Fluid

The several tasks the automatic transmission fluid must perform are:

Power Transmission. The fluid in an automatic transmission is used as the medium for transmitting the engine's power to the rear wheels. There is no mechanical connection between the engine and the power train. The connection is made by fluid flow in the torque converter.

Pressure Transmission. Fluids are used to sense the vehicle's transmission requirements. Fluids under pressure are directed to components in the transmission which sense and select the correct gear ratio for the vehicle's operating requirements.

Lubrication. The same fluid which transmits the engine's power and provides pressure to operate the transmission must also lubricate the transmission's components. The fluid is circulated in the transmission to provide lubrication for all the bearings, bushings, seals, and gears.

Cooling. The fluid in an automatic transmission provides the means of removing the heat from the transmission. Like the water circulated around engine parts to remove heat, fluid is circulated around parts in the automatic transmission to remove heat.

The Properties of Automatic Transmission Fluid

The automatic transmission fluid must serve to transmit pressure and power, and to cool and lubricate. It must do these tasks over a temperature range of from $-40°$F to an excessive $+350°$F and under a wide variety of service conditions. Several properties or characteristics must be present in the fluid to allow it to accomplish these functions. The more important properties are described below.

Viscosity. Viscosity describes the thickness or thinness of a fluid. A common term for viscosity is "body." A fluid with a high viscosity is said to have a heavy body. An example of a fluid with a high viscosity is the thick lubricant used in standard transmissions and rear axles. A low viscosity fluid is one with a light body. The viscosity of a fluid determines how freely it flows. Low viscosity fluids flow very freely; high viscosity fluids flow sluggishly. Automatic transmission fluids must have a viscosity which allows them to flow freely in cold conditions but which will allow them to have sufficient body during times of high temperature.

Viscosity Index. A fluid's viscosity changes as the temperature changes. Pancake syrup is thick at room temperature but when heated it becomes thinner.

Its viscosity has changed from high to low. A fluid which changes a good deal with temperature changes is said to have a low viscosity index. A fluid which changes relatively little with a given temperature change is said to have a high viscosity index. Fluids for automatic transmissions must have a high viscosity index since these transmissions must operate in the same way under both very hot and very cold conditions.

Lubricity. Lubricity refers to how well a fluid works to prevent wear between closely fitted working parts. Direct metal-to-metal contact is avoided by forcing the fluid between the parts. If the fluid has sufficient lubricity it will stay in place and prevent wear.

Foam Resistance. Automatic transmission fluid must resist foaming. The rotating components in the transmission act as an eggbeater to constantly churn the fluid. If a foam were to form, it would interfere with the fluid's ability to lubricate and transmit pressure.

Oxidation Resistance. Oxidation occurs when a fluid combines chemically with oxygen. The three greatest factors that promote oxidation are air, heat, and contamination. All three are present in an automatic transmission. If an excessive amount of oxidation occurs, a sludge begins to form on components. This sludge deposit may interfere with the operation of the transmission.

Anti-rust. Water and oxygen can cause rusting of the steel and cast iron components inside an automatic transmission. Air inside the transmission can provide both oxygen and water. Rusting results in surface destruction and rust may be carried into the hydraulic system where it will contribute to the formation of sludge.

Compatibility. Fluids for use in automatic transmissions must be compatible with the metallic and non-metallic parts used in the transmission.

Fluid Composition

An automatic transmission fluid is probably one of the most complicated lubricants formulated due to the variety of functions it must serve. The fluid is made up of a number of components, beginning with an oil called a base oil. To the base oil a second component called viscosity improver is added. The viscosity improver helps the oil resist changes in its viscosity as temperatures change. An additive package is then combined with this mixture. The additive package is the group of chemicals which give the fluid its desirable qualities. Typical additives are oxidation inhibitors, extreme pressure agents, detergents, dispersants, and seal swell agents. To this mixture a red dye is added to make the fluid easy to recognize from other fluids used in the vehicle.

Types of Fluids for Automatic Transmissions

Several types of fluids are manufactured for use in automatic transmissions. These fluids are identified by letters printed on the container. The letters and their interpretations are as follows:

ATF. The letters "ATF" printed on the container stand for Automatic Transmission Fluid. This is the standard abbreviation used by oil companies and transmission manufacturers.

Type A. Engine oils were used to lubricate the earliest automatic transmissions. These oils proved inadequate for the job they had to do. In the late 1940s a program was initiated to develop a set of specifications for suitable fluids. This set of specifications spelled out such things as viscosity, anti-foam properties, oxidation limits, frictional properties, etc. These specifications were named "Type A" specifications. A company that manufactured automatic transmission fluid to these specifications was allowed to print "Type A" on the container. Fluid so designated was labeled "ATF-Type A."

Suffix A. In the late 1950s, deficiencies in oxidation resistance were found in some "Type A" fluids. A major specification change was undertaken. Fluids meeting the new specifications were designated "Type A, Suffix A." Fluid manufacturers who made fluid to these new specifications printed "Type A, Suffix A" on their containers. The "Type A, Suffix A" specifications remained unchanged until the 1960s.

Type F. Prior to 1961, the Ford Motor Company specified "Type A, Suffix A" fluid for its automatic transmissions. Increased performance requirements led the Ford engineers to develop a more stringent set of specifications for automatic transmission fluid. These specifications described new viscosity, antiwear, friction, and oxidation requirements. Fluid which is manufactured to Ford's specifications is described as "Type F" fluid. The complete designation on the container is "ATF-Type F."

Dexron. General Motors Corporation also felt the need to develop a new set of specifications for their increasing service requirements. Engine horsepower had increased rapidly, making the transmission's job more difficult. The new specifications were designed to overcome these difficulties. General Motors labeled these specifications "Dexron." "Dexron" is a coined word registered as a trademark by General Motors. Each supplier of a qualified fluid is licensed to use the mark "Dexron" to identify the fluid.

The correct type of fluid must always be used in an automatic transmission. The vehicle manufacturers decide which set of specifications is most appropriate for their transmissions. The type of automatic transmission fluid which meets these specifications is recommended for the vehicle. The vehicle owner's manual and service manual will both list what type of fluid should be used. The use of an incorrect type when adding or filling can result in damage to the transmission. If fluid used is under the specifications required, the transmission will not be properly protected from wear. If the fluid is not the correct type, there is the added danger that the fluid will not be compatible with the non-metallic material used in the transmission.

UNIT 2
Fluids for Automatic Transmissions

True-False Questions (Indicate the correct choice)

1. Fluid in an automatic transmission is used as the means of transmitting the engine's power to the power train and thus to the rear wheels. T F

2. One task of the fluid is to lubricate the transmission's components. T F

3. The term "viscosity" describes how thick or thin a fluid is. T F

4. Low viscosity fluids flow sluggishly. T F

5. As a fluid is heated its viscosity gets higher. T F

6. Automatic transmission fluid must resist foaming. T F

7. Oxidation of automatic transmission fluid is desirable. T F

8. The letters "ATF" printed on a container stand for Automatic Transmission Fluid. T F

9. The first fluids manufactured specifically for automatic transmissions were called "Type F" fluids. T F

10. A major specification change to improve the oxidation resistance of "Type A" fluids resulted in "Type A, Suffix A" fluids. T F

11. Increased performance requirements led the Ford Motor Company to develop more stringent specifications for ATF, resulting in "Type F" fluids. T F

12. "Dexron" is a term used to describe ATF which was used in the earliest automatic transmissions. T F

13. Only the type of fluid recommended by the manufacturer should be used in an automatic transmission. T F

14. If an incorrect type of fluid is used in an automatic transmission, the transmission may not be properly protected from wear and it may not operate correctly. T F

15. All ATF is compatible with each manufacturer's transmission components. T F

Multiple Choice Questions (Indicate the correct choice)

1. The function of the fluid in an automatic transmission is:
   a. to transmit power
   b. to transmit pressure
   c. to provide lubrication
   d. to cool the transmission
   e. all of the above

2. The amount a fluid changes viscosity as temperature changes is its:
   a. Viscosity Index
   b. Variability Index
   c. Temperature Index

3. The term used to describe how well a fluid works to prevent wear is:
   a. viscosity
   b. compatibility
   c. lubricity
   d. oxidation

4. Which of the following is *not* one of the major components of automatic transmission fluid?
   a. oxidation inhibitors
   b. base oil
   c. viscosity improver
   d. additive package

UNIT 2
Fluids for Automatic Transmissions

True-False Questions

1. This statement is true. The connection between the engine and the power train is made by fluid flow in the torque converter.

2. This statement is true. The fluid is circulated in the transmission to provide lubrication for all the bearings, bushings, seals, and gears.

3. This statement is true. Viscosity describes the thickness or thinness of a fluid.

4. This statement is false. Low viscosity fluids flow very freely. High viscosity fluids flow sluggishly.

5. This statement is false. A fluid's viscosity changes as temperature changes. As it is heated it gets thinner or has a lower viscosity, rather than a higher viscosity.

6. This statement is true. If a foam were to form in an automatic transmission fluid, it would interfere with the fluid's ability to lubricate and transmit pressure.

7. This statement is false. If an excessive amount of oxidation of automatic transmission fluid occurs, a sludge begins to form on components. This sludge deposit may interfere with the operation of the transmission.

8. This statement is true. ATF is the standard abbreviation used for automatic transmission fluid.

9. This statement is false. The first specifications for automatic transmission fluids resulted in "Type A" fluids (not "Type F").

10. This statement is true. "Type A, Suffix A" fluids were an improvement over "Type A" fluids, brought about by correction of deficiencies in oxidation resistance.

11. This statement is true. "Type F" fluids were developed to meet increased performance requirements.

12. This statement is false. "Dexron" is a term used to describe the ATF developed by General Motors as an improvement over "Type A, Suffix A" fluids.

13. This statement is true. The recommended type of fluid must always be used in an automatic transmission.

14. This statement is true. If the fluid used is under the specifications required, the transmission will not be properly protected from wear and may not operate correctly.

15. This statement is false. All ATF is not compatible. If the wrong fluid is used,

there is a danger that it will not be compatible with the non-metallic components of the transmission.

Multiple Choice Questions

1. The correct choice is *e*. The fluid must serve to transmit pressure and power, and to cool and lubricate.

2. The correct choice is *a*. The Viscosity Index describes how much a fluid changes viscosity with temperature change.

3. The correct choice is *c*. Lubricity refers to how well a fluid works to prevent wear between working parts.

4. The correct choice is *a*. The major components of automatic transmission fluid are base oil, viscosity improver, and additive package. Oxidation inhibitors are part of the additive package.

Section II

FLUID COUPLINGS AND TORQUE CONVERTERS

Fluid Couplings

The first major section of the automatic transmission is the fluid coupling or torque converter. Early automatic transmissions use a fluid coupling. Later model transmissions use a torque converter. The fluid coupling's job is roughly similar to that of the clutch in front of a standard transmission. It serves to connect and disconnect the engine from the power train so that the vehicle can be cranked for starting and so it may run (idle) without moving. How the fluid coupling works is presented in this unit.

Energy in a Stream of Fluid

The idea of using fluid energy was not new when the first fluid coupling was invented. The energy in a stream of fluid had been harnessed for a long time. An example is the ordinary water wheel shown in figure 3-1. A flow of water from a fast moving stream or waterfall was directed onto the blades of the water wheel. If the wheel was mounted so that it was free to turn, the force of the water against the blades would cause it to turn. The faster the water flow directed against the wheel, the faster it would turn. The rotating wheel could then be used to drive various kinds of machinery. The idea of directing a stream of fluid at a wheel with blades is used in the fluid coupling.

Figure 3-1
Waterwheel Using the Energy in a Stream

Components of a Fluid Coupling

The working components of a fluid coupling look a good deal like a doughnut which is sliced down the middle so there is no connection between the two halves. One half is fastened to the

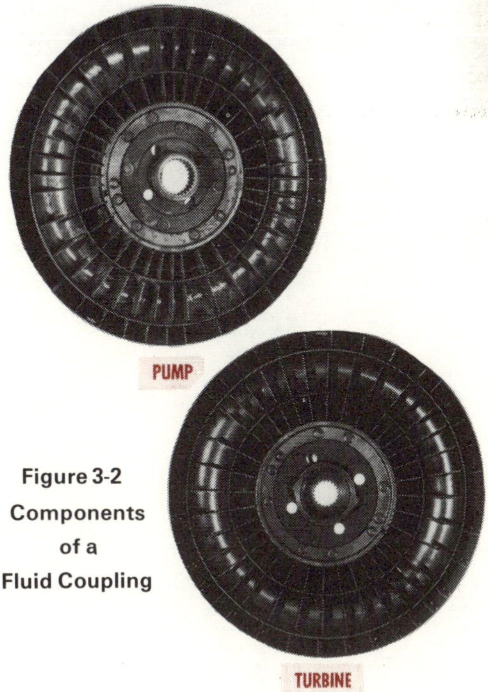

Figure 3-2
Components of a Fluid Coupling

engine's crankshaft. This half is called the *pump*. The other half is attached to the input shaft of the planetary gearbox. This half is known as the *turbine*. The unit is hollow, but each half has a number of straight radial blades leading from the hub to the outside edge. The coupling will most often have a cover around it which is fastened solidly to one of the rotating members. The hollow space between the pump and turbine is filled with fluid.

Operation of the Fluid Coupling

When the engine is started, the fluid coupling pump, which is mounted to the engine crankshaft, begins to rotate. As the pump turns, fluid is thrown toward its outside edge by centrifugal force. When it gets to the outside edge it wants to keep going, but the only place it can go is across to the other half of the coupling—the turbine. All the time that the fluid is being forced outward, it is also being whirled around in the direction of engine rotation by the pump blades. Consequently, when the fluid crosses over into the turbine, it hits the turbine blades and pushes the turbine around in the same way that the water wheel was forced to turn.

The principle of the fluid coupling can be illustrated with two fans. One fan (A) is plugged into an electrical outlet. Another fan (B) is not connected to an electrical outlet but is positioned directly in front of fan A. When fan A is turned on, air is thrown off its blades toward the blades of fan B. This causes the blades of fan B to turn. In this example, fan A could be considered the pump (connected to the engine); fan B, the turbine (connected to the planetary gearbox); and air, the fluid.

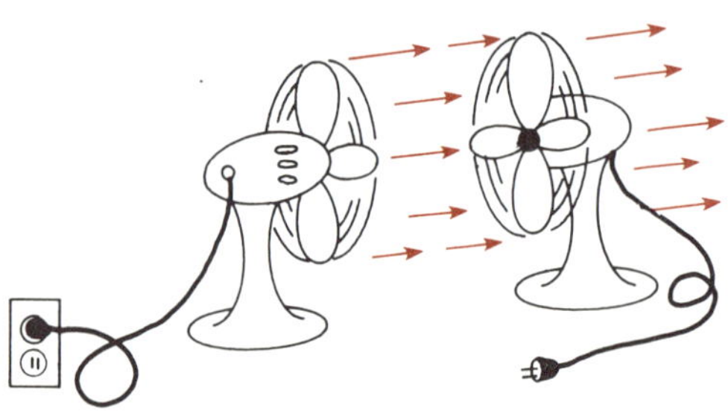

Figure 3-3

Fan Analogy of a Fluid Coupling

After the fluid has pushed the turbine blades it tends to slow down, and then it travels toward the center, or hub, of the turbine. From there it crosses back to the center of the pump, and the whole process is repeated. Thus, the fluid is continually circulated outward in the pump, over to the turbine, into the center of the turbine, and back over to the pump.

The fluid coupling acts in much the same way as the clutch on a standard transmission. The difference is that it does so automatically. The engine can be crank-started only if it is disconnected from the power train. On a standard transmission car, the clutch is used to uncouple the engine from the power train for starting. The fluid coupling does this job automatically, simply because at cranking speed fluid cannot be thrown hard enough from the pump to cause the turbine to rotate. The vehicle can idle without moving for the same reason: Fluid is not thrown hard enough at the turbine at low engine speeds to make the vehicle move. Only when the accelerator pedal is depressed to make the engine turn faster than at idle does the fluid strike the turbine with sufficient force to move the vehicle.

Path of Fluid Flow

The path of fluid flow in the fluid coupling is described by the term "vortex." Vortex describes the movement of fluid from the pump to the turbine and back to the pump. The path of the fluid is affected by the fact that the pump and turbine seldom rotate at the same speed. Consider the situation that exists when the vehicle's engine is running, but its rear wheels have not yet begun to turn. The turbine is stationary while the pump is rotating at perhaps 1,000 rpm. The fluid is thrown off the pump blades to the turbine. By the time the fluid runs to the center of the turbine and reenters the pump, the pump has rotated quite some distance. Consequently, it is impossible for the fluid to return to the same row of pump fins from which it left. If a single drop of fluid were examined as it circulated through the fluid coupling, its path, or vortex, would look like that shown in figure 3-4.

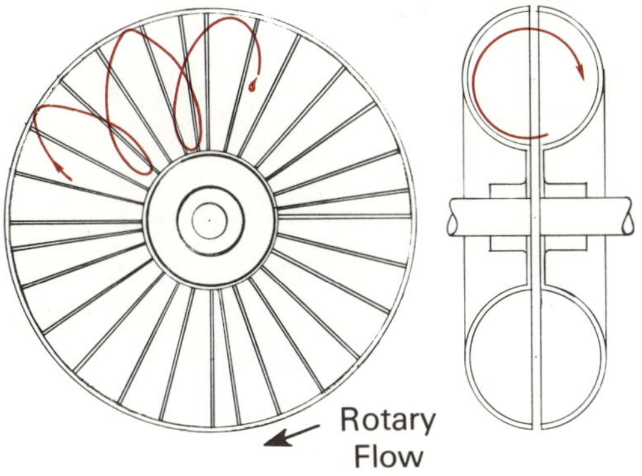

Figure 3-4

Vortex Flow in a Fluid Coupling

The forces which influence this path of fluid flow are the rotary action of the pump, the centrifugal force resulting from the rotation of the pump, the shape of the pump blades, the shape of the turbine blades, and the difference in speed between the pump and turbine. It should be pointed out that we have been examining one vortex. Obviously there are as many vortex streams in a fluid coupling as there are fins or blades on the pump and turbine.

High and Low Vortex

When the difference in speed between the pump and turbine is great, as is the case when the vehicle has not yet started to move, a condition of "high vortex" is said to exist. As the vehicle begins to move and pick up speed, pump and turbine begin to turn at nearly the same speed. The pump will not turn very far before the fluid is returned to its inner row of fins from the turbine. Under these conditions the coils of the vortex flow would be closer together as shown in figure 3-5. This is a condition of "low

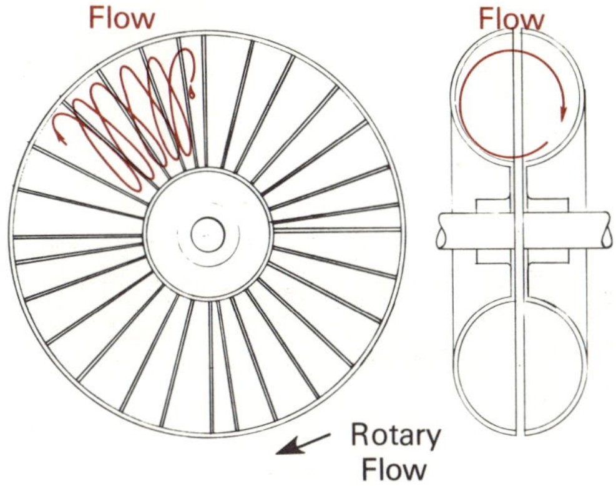

Figure 3-5

Low Vortex

vortex." The degree of vortex is constantly changing, as the vehicle and engine speed affect the difference in speed between the pump and turbine. The higher the vortex, the greater the driving power of the fluid. When the vortex is high, the fluid stream strikes the blades of the turbine at nearly a right angle. As the vortex decreases, the angle at which the fluid strikes the blades becomes less and less favorable, as shown in figure 3-6A and 3-6B (next page).

Zero Vortex or Coupling Stage

Zero vortex or "coupling stage" is a condition in which the pump and turbine are turning at exactly the same speed.

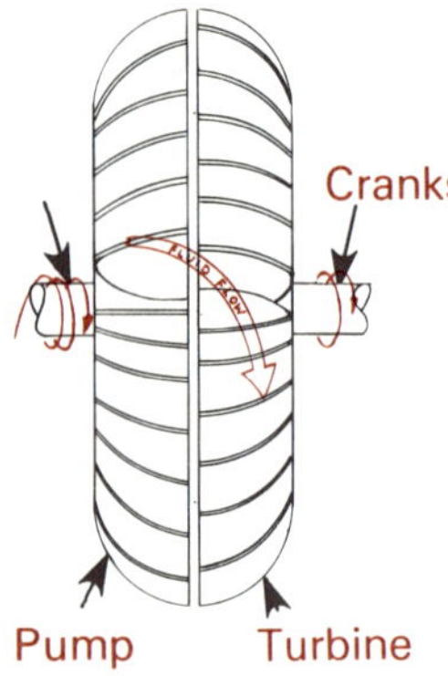

Figure 3-6A

High Vortex

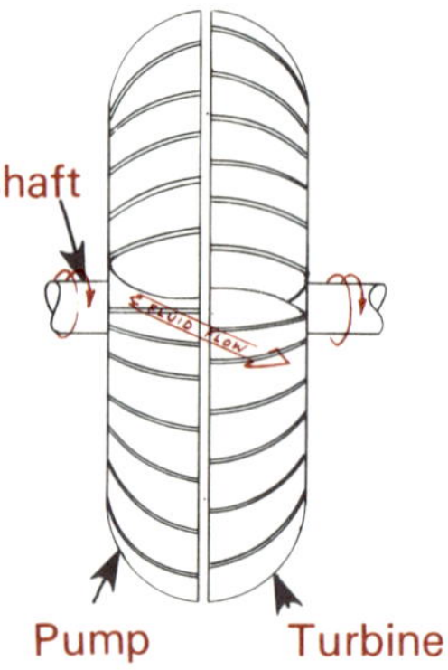

Figure 3-6B

Low Vortex

Fluid is not thrown from one unit to another, it simply rotates *with* the pump and turbine. The only flow is rotary flow as shown in figure 3-7. Since there is no vortex stream hitting the turbine blades

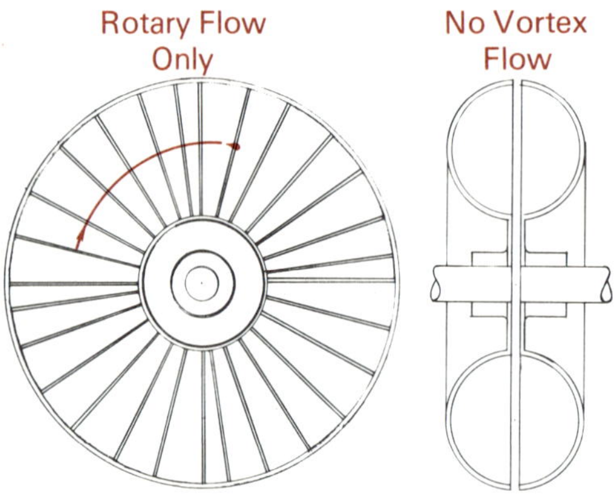

Figure 3-7

Zero Vortex

there would be no force applied to the turbine. Such a condition seldom exists in reality, however, because the turbine always lags just a little behind the pump. A true fluid coupling stage occurs only when the vehicle begins to coast, but even then not for very long, because as soon as the engine slows down, a reverse vortex is established. Fluid is thrown from the turbine to the pump and the engine is able to provide compression braking.

Application

The fluid coupling has been used on many automatic transmissions. Until the 1950s, almost all automatic transmissions used one. In the 1950s, many manufacturers began to drop this unit in favor of a torque converter. The fluid coupling, however, is still used in a great many applications other than automobiles. They are used in the power transmission systems of generators, ships, dynamometers, and many other devices.

Advantages of the Fluid Coupling

Wherever fluid couplings are used, they provide several advantages over a mechanical coupling. One advantage, already discussed, is that coupling and uncoupling of the engine from the power train is done automatically. Another advantage is that shock loads are not transmitted from the engine to the power train or from the power train to the engine. The fluid coupling acts as a shock absorber to protect both engine and power train against damage. The slippage which occurs in a fluid coupling will also protect the engine from lugging or running at too low a speed. A good deal of harm can occur to an engine which is forced to run at too low a speed. When the load is too great for the engine, the coupling will slip, allowing the engine to maintain its speed. Finally, the fluid coupling provides for a much smoother flow of power to the rear wheels than is possible with a mechanical coupling.

UNIT 3
Fluid Couplings

True-False Questions (Indicate the correct choice)

1. The fluid coupling's job is roughly similar to the manual clutch in front of a standard transmission. T F

2. The idea of using fluid energy was first used in the fluid coupling. T F

3. The fluid coupling pump is connected to the engine's crankshaft. T F

4. The fluid coupling turbine is attached to the input shaft of the planetary gearbox. T F

5. The path of fluid flow in the fluid coupling is described by the term "vortex." T F

6. When the pump and turbine begin to turn at nearly the same speed, the fluid path is described as "high vortex." T F

7. When the difference in speed between the pump and turbine is great, a condition of "high vortex" exists. T F

8. The lower the vortex, the greater the driving power of the fluid. T F

9. Zero vortex is a condition in which the pump and turbine are turning at exactly the same speed. T F

10. The fluid coupling provides for a much smoother flow of power to the rear wheels than is possible with a mechanical coupling. T F

Multiple Choice Questions (Indicate the correct choice)

1. Which of the following is *not* a component of the fluid coupling?
   a. pump b. turbine c. stator

2. The force(s) which influence vortex flow are:
   a. Rotary action of the pump.
   b. Centrifugal force resulting from pump rotation.
   c. Shape of pump and turbine blades.
   d. Difference in speed between pump and turbine.
   e. All the above.

3. When the difference in speed between the pump and turbine is great, as when the vehicle has not yet started to move, the vortex is:
   a. high b. low c. zero

4. When the vehicle is traveling down the highway with pump and turbine turning at nearly the same speed, the vortex is:
   a. high b. low c. zero

5. The greatest driving power is provided by:
   a. high vortex b. low vortex c. zero vortex

6. Which of the following is *not* one of the advantages of a fluid coupling?
   a. Coupling and uncoupling is achieved automatically.
   b. Shock loads are not transmitted from the engine to the power train.
   c. Torque is multiplied.
   d. The engine is protected from lugging.

Parts Identification and Fluid Path

Identify the components of a fluid coupling by writing in their names in the spaces provided. Draw in the direction of fluid flow.

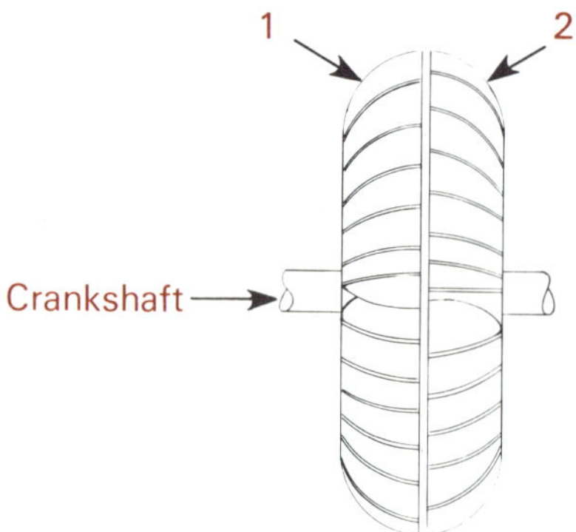

1.____________________

2.____________________

UNIT 3
Fluid Couplings

True-False Questions

1. This statement is true. The fluid coupling connects and disconnects the engine from the power train so that the vehicle can be cranked for starting and so that it may idle without moving.

2. This statement is false. The energy in a stream of fluid had been harnessed for a long time in water wheels.

3. This statement is true. The fluid coupling pump is mounted on and driven by the engine's crankshaft.

4. This statement is true. The fluid coupling turbine is attached to the input shaft of the planetary gearbox.

5. This statement is true. "Vortex" describes the movement of fluid from the pump to the turbine and back to the pump.

6. This statement is false. When the pump and turbine begin to turn at nearly the same speed, a condition of "low vortex" exists (not "high vortex").

7. This statement is true. "High vortex" occurs when the difference in speed between the pump and turbine is great.

8. This statement is false. When vortex is ***high*** the fluid stream strikes the blades of the turbine at nearly a right angle, imparting a strong driving power.

9. This statement is true. Zero vortex, or "coupling stage," is a condition in which the pump and turbine are turning at exactly the same speed. Fluid is not thrown from one unit to another, but simply rotates with the pump and turbine.

10. This statement is true. An important advantage of a fluid coupling is the smooth flow of power it provides.

Multiple Choice Questions

1. The correct choice is *c.* The fluid coupling has two components—a pump and a turbine.

2. The correct choice is *e.* The path of fluid flow is influenced by the rotary action of the pump, the centrifugal force resulting from pump rotation, the shape of the pump and turbine blades, and the difference in speed between pump and turbine.

3. The correct choice is *a.* When the difference in speed of the pump and turbine is

great, such as when the vehicle has not yet started to move, a condition of "high vortex" exists.

4. The correct choice is *b.* When the pump and turbine speeds are nearly the same, the vortex is low.

5. The correct choice is *a.* When vortex is high, the fluid stream strikes the blades of the turbine at nearly a right angle, providing a strong driving power.

6. The correct choice is *c.* The fluid coupling has many advantages over a mechanical coupling but it *cannot* multiply torque.

Parts Identification and Fluid Path

The correct names of the components and the path of fluid flow are shown below.

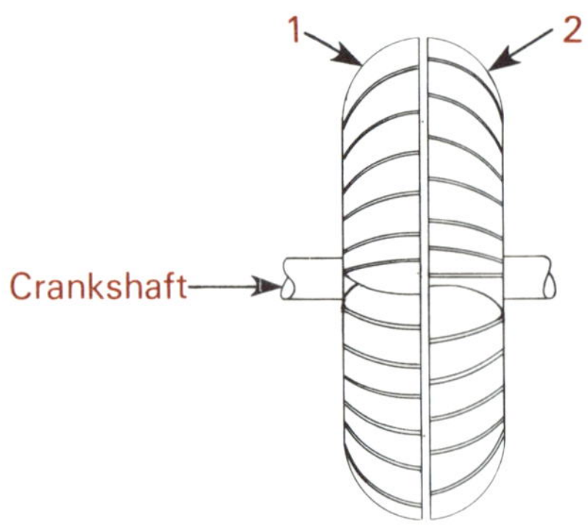

1. Pump

2. Turbine

Torque Converters

Most automatic transmissions today use a torque converter. The torque converter serves the same function as the fluid coupling on earlier transmissions; that is, it serves to couple and uncouple the engine from the power train. The torque converter does so with all the advantages of a fluid coupling. The converter has one important advantage over a fluid coupling—it can multiply torque.

Torque

Torque is a word used to describe a turning or twisting effort applied to something to make it turn, or try to make it turn. The mechanic who uses a wrench to tighten a nut or bolt is exerting a torque on the nut or bolt through the wrench, as shown in figure 4-1.

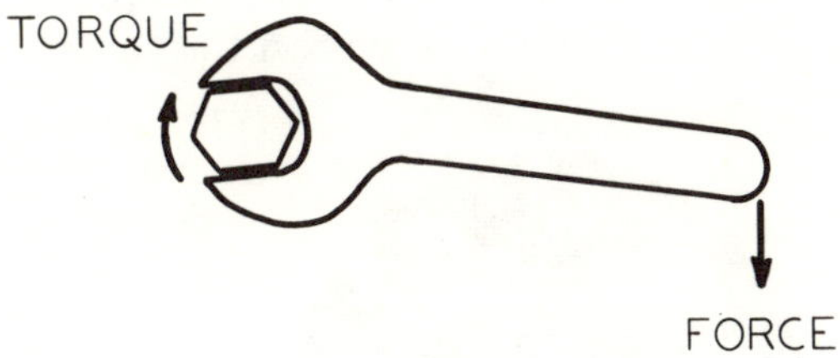

Figure 4-1

Torque Is Turning Effort

As every mechanic knows, if a longer wrench is used, more force, or torque, can be applied to the bolt. The standard measure of torque is the pound-foot. Torque is calculated by multiplying the pounds of force applied (to the wrench, for example) times the length of the lever (length of the wrench). In the example shown in figure 4-2, two wrenches of different lengths are used to tighten a bolt. If a 20-pound force is applied to a wrench that is one foot long, the torque

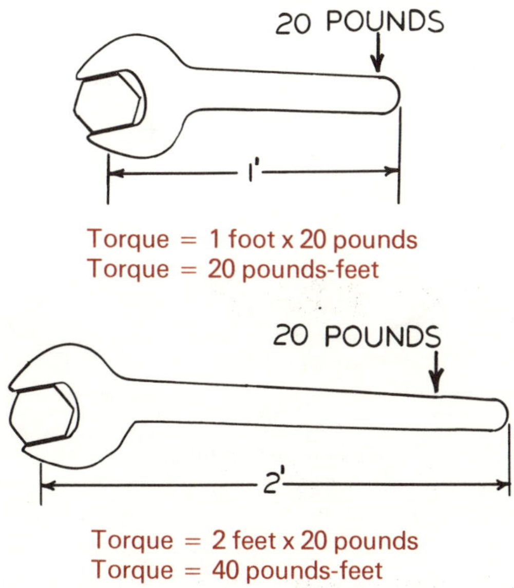

Figure 4-2

Torque Equals Force Times Lever

applied to the bolt is 20 pounds-feet (1 foot × 20 pounds = 20 pounds-feet). If a wrench twice as long were used with the same force, 40 pounds-feet would be exerted on the bolt (2 feet × 20 pounds = 40 pounds-feet).

Torque Multiplication

Torque is used in an automobile to rotate the rear wheels. This torque is received from a rotating crankshaft. When a vehicle is rolling on level ground, the engine develops enough torque to keep it moving. On a hill, or when moving the vehicle from a standstill, the driving wheels need more torque. The engine's torque must be multiplied. The common way of multiplying torque is through gears.

Assume that an engine were to deliver 200 pounds-feet of torque to a gear with a one-foot radius which is meshed to another gear with a two-foot radius (see figure 4-3). The torque applied to the second gear would be 400 pounds-feet, or twice the engine torque. Torque is multiplied because the "lever" on the second gear is twice the length of the "lever" on the first gear. Thus, just as torque can be increased with a longer wrench, it can be multiplied with larger gears. It should be pointed out, however, that when torque is multiplied, speed is reduced. In the example above, the large gear turns more slowly than the small gear. Torque is gained at the expense of a loss in output rpm. Stated another way, torque increase means speed reduction. If torque is doubled, for example, speed is divided in half. Gears are not the only way torque can be multiplied. Torque multiplication can be achieved hydraulically in a device called a torque converter.

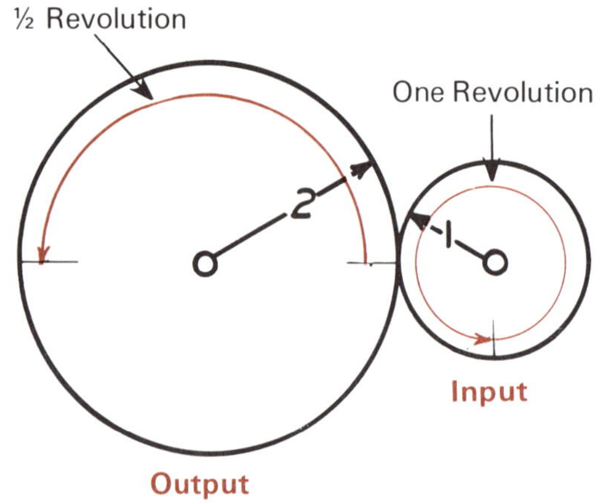

Figure 4-3

Engine Torque Multiplied by Two

Components of a Torque Converter

Like the fluid coupling, the torque converter has a pump and a turbine. The pump is connected to the engine and the turbine is connected to the input shaft of the transmission. The blades on the torque converter pump and turbine are curved, while those on the fluid coupling are straight, as shown in figure 4-4. The curvature of the blades is designed to move the fluid as efficiently as possible.

The torque converter has a third component, not present in a fluid coupling. This component is called a stator. The stator fits between the pump and turbine. Its job is to redirect the vortex to

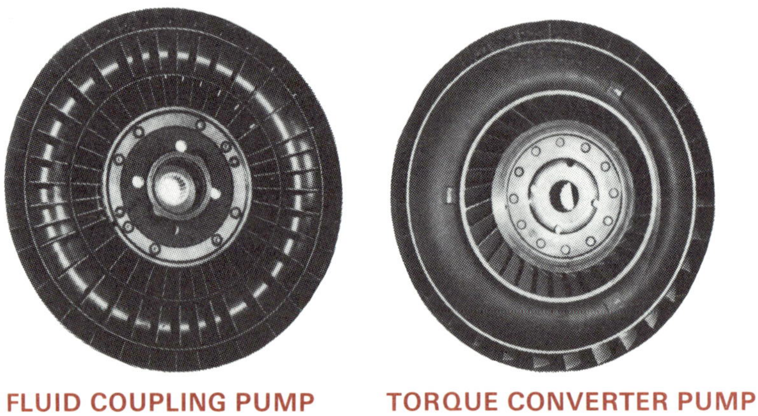

Figure 4-4

Curved Blades Are Used on a Torque Converter

Figure 4-5

Components of a Torque Converter

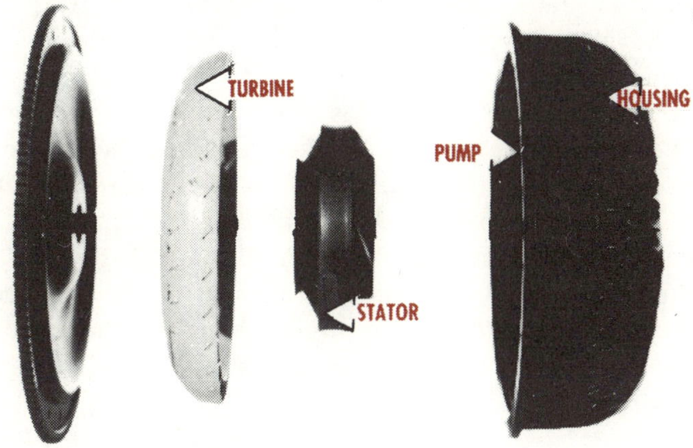

help the pump turn rather than opposing it. The components of a torque converter are shown in figure 4-5. The units are enclosed in a housing and the entire housing is filled with fluid. The converter housing is bolted together on earlier model transmissions, and welded together on later model transmissions. Thrust washers are used to take up the thrust between the units, and between the turbine and the housing. The entire unit is then bolted to the flywheel of the engine.

Operation of the Torque Converter

When the engine is started, the pump (sometimes called the impeller) begins to rotate. Fluid is thrown outward along the pump blades by centrifugal force. The fluid has nowhere to go when it reaches the outside edge of the pump except over to the turbine, where it then pushes on the turbine blades. If the engine is turning fast enough, fluid will strike the turbine with sufficient force to turn it. Unlike the fluid coupling, the fluid's job is not complete here. The blades on the turbine are curved to direct the fluid into its center and toward the stator.

The stator has blades curved in such a way that they change the direction of the fluid as it leaves the turbine. The fluid is then directed to the pump, where it strikes the pump in the direction of its rotation. The energy in the vortex is thus directed back from turbine to pump, and is used to assist the engine in driving the pump, as shown in figure 4-6.

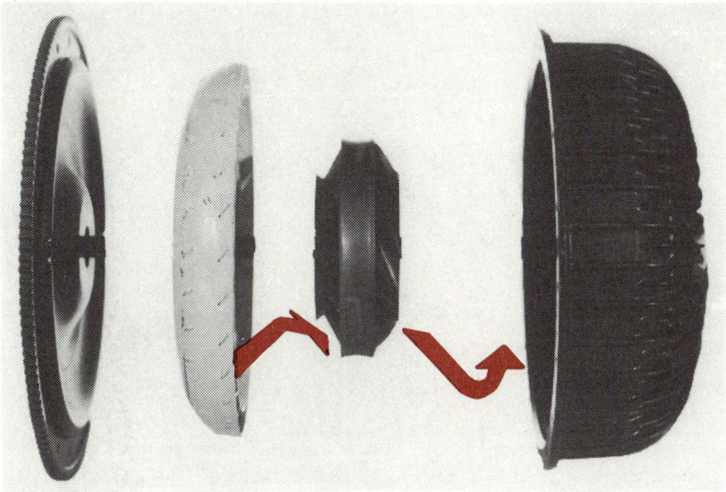

Figure 4-6

Stator Operation

The two fans may be used once again to provide a simple illustration of the torque converter principle. When fan A (the pump) turns, its blades throw air (fluid) at fan B (the turbine). When fan A turns fast enough the force of the air flow will cause fan B to turn. Up to this point the operation is exactly like the fluid coupling. In this case, however, a device called a stator is added to redirect the air from fan B back to the blades of fan A.

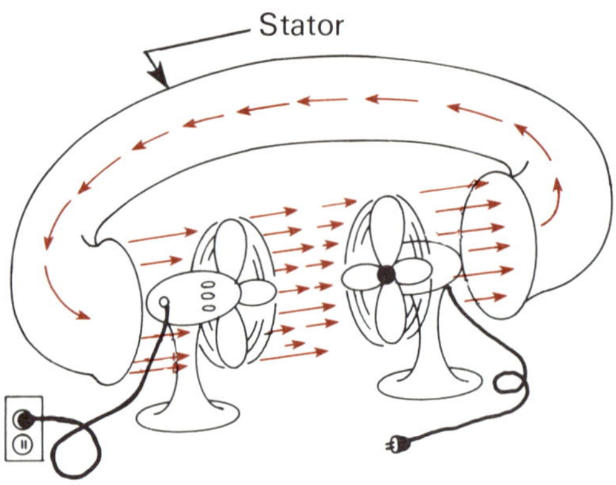

Figure 4-7

Fan Analogy of a Torque Converter

This redirected air helps the blades of fan A turn faster than they normally would if powered only by the fan's motor. Thus, the torque or turning effort of the fan has been increased or multiplied.

The stator's job is to provide some directional control to the fluid as it leaves the turbine and enters the pump. It is needed only during times of high vortex, or when the pump and turbine are turning at different speeds. The stator is not needed at low vortex because the pump and turbine are turning at nearly the same speed and there is very little vortex flow to be redirected. For this reason, the stator is mounted on a shaft called a "reaction shaft." This shaft does not rotate, but is mounted to the front of the transmission and protrudes into the torque converter. The stator is attached to the reaction shaft through a one-way clutch. The one-way clutch consists of a number of spring-loaded rollers which allow the stator to turn in one direction, but prevent it from turning in the other direction. The stator clutch is shown in figure 4-8.

The stator is locked or not locked to the reaction shaft according to vortex conditions. During conditions of high vortex, the fluid leaves the turbine at an angle which enables it to strike the stator squarely on the front of its blades. Pushing the stator in this direction locks up the one-way clutch. The stator is sta-

Figure 4-8

Stator Clutch

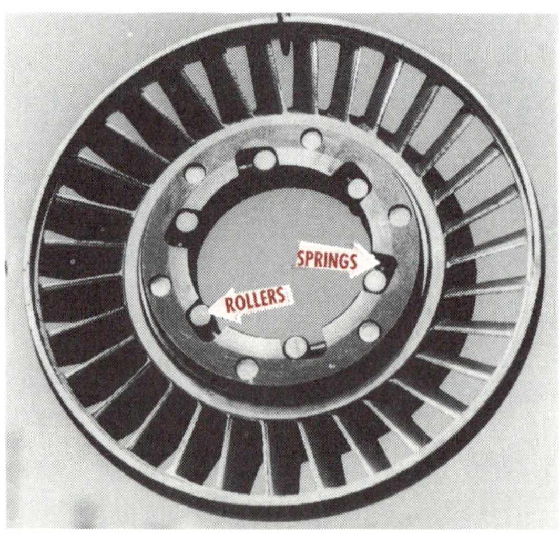

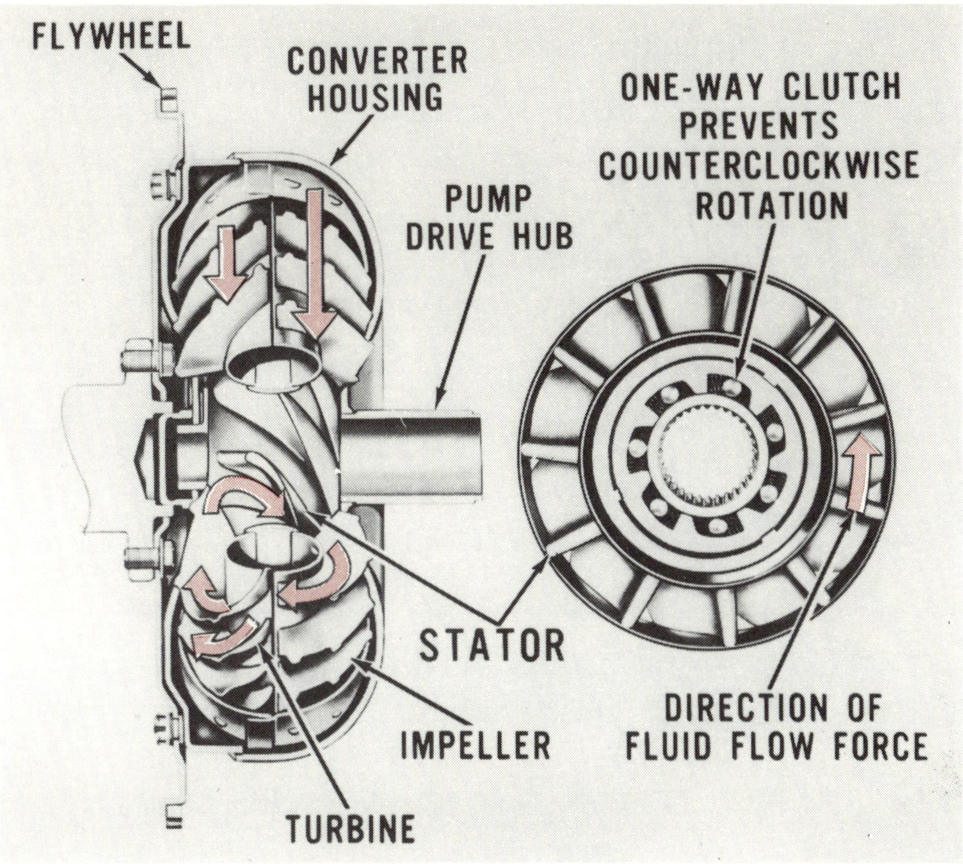

Figure 4-9

Fluid Hits Stator to Lock It at High Vortex
(Ford Motor Co.)

tionary and able to redirect oil back into the pump, as shown in figure 4-9.

As the vehicle begins to come up to cruising speed, a condition of low vortex exists. The fluid leaves the turbine and enters the stator in a different direction, so that it strikes the back of the stator blades. This is the direction in which the stator clutch permits rotation, so the stator will now rotate in the same direction as the pump and turbine. Thus, as the turbine speed approaches pump speed, the stator "free wheels" and is carried along with the rotating fluid mass, as shown in figure 4-10. If the vortex were to change back to high, as would be the case if the driver accelerated, the stator would again be locked up and would provide fluid redirection.

Torque Conversion

A torque converter can multiply torque while a fluid coupling cannot. In the case of a vehicle equipped with a fluid coupling, starting the vehicle from a standstill is very similar to starting a car with a geared transmission in high gear and allowing the clutch to slip in order to get off to a smooth start. As the vehicle begins to move, with the engine rotating at about 1,500 rpm, the drive shaft turns very slowly; then it picks up speed until finally it rotates at almost the same speed as the crankshaft. There is no torque multiplication. The same amount of

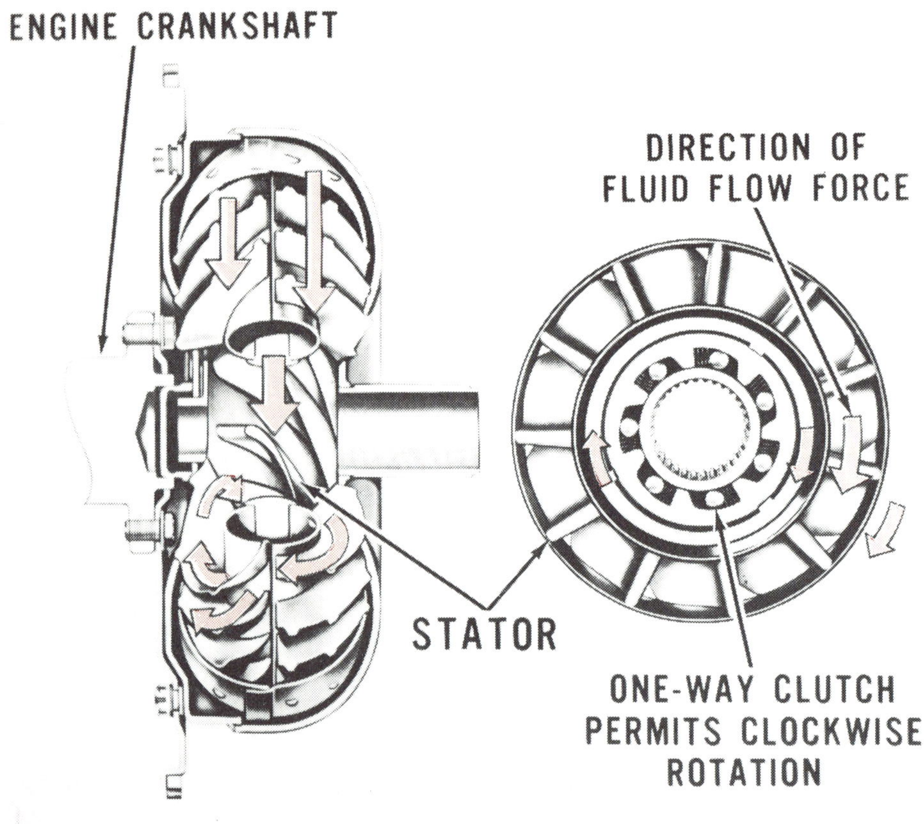

Figure 4-10

Fluid Hits Stator to Unlock It at Low Vortex
(Ford Motor Co.)

torque comes out of the fluid coupling as went into it. Acceleration from a standstill is slow. Although the fluid coupling does a better job than a slipping clutch, and has many advantages over a mechanical coupling, all the multiplication of engine torque must be accomplished in the gearbox.

The torque converter minimizes the necessity of using so many gears to get the vehicle in motion. Engine torque which enters the torque converter can be increased as it goes out. If a vehicle with a torque converter starts from a standstill with the engine at 1,500 rpm and with the drive shaft just beginning to move, the effect through the torque converter is the same as if the engine were turning the drive shaft through gears which would afford torque multiplication. As the vehicle begins to move, the degree of torque multiplication decreases until the coupling state is reached, and there is no longer any torque multiplication.

The torque converter multiplies engine torque in proportion to the difference in speeds between the pump and turbine. When vortex is high, there is a good deal of torque multiplication. When vortex is low, there is very little torque multiplication. At zero vortex there is no torque multiplication. This, of course, is exactly what is required, because at low rpm engine torque is low and must be increased to get the vehicle moving.

Torque is able to be multiplied in a

torque converter because stationary blades on the stator redirect fluid back to the pump to help it turn. In figure 4-11, if a vortex is directed at a bucket attached to the rim of a wheel the wheel will turn. This compares to the fluid being thrown from the pump blades to the turbine blades in the torque converter. The fluid swirls around the curved surface of the bucket and leaves the bucket with almost the same velocity it had when entering. The fluid will give up very little of its energy to the bucket and there will be only a small push on the bucket. Similarly, there would be a light push or torque applied to the converter turbine.

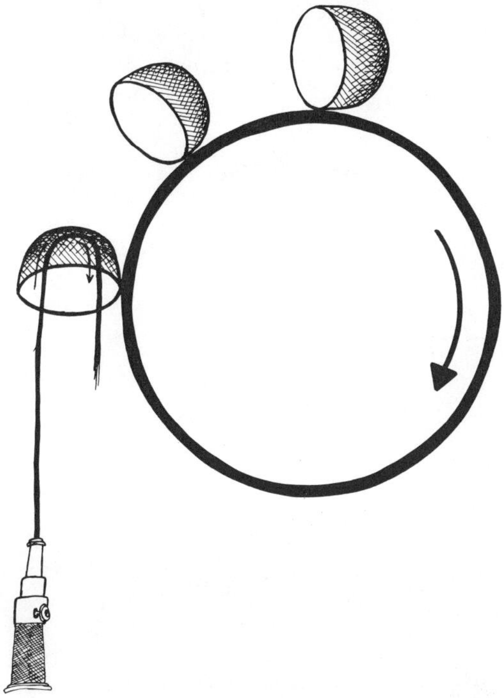

Figure 4-11

Illustration of a Small Amount of Torque Applied to Wheel

If a stationary curved vane is installed, as shown in figure 4-12, fluid which comes out of the bucket will enter the stationary curved vane where it will reverse direction and reenter the bucket.

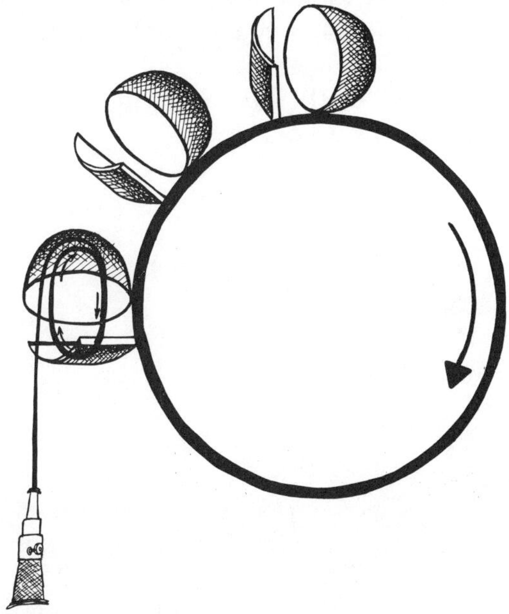

Figure 4-12

Illustration of a Multiplying Torque on Wheel

It is theoretically possible for the fluid to make many complete trips between the vane and bucket. Each time the fluid reenters the bucket a force or torque is exerted on the bucket. This is torque multiplication, or torque conversion. The stationary vane in the torque converter is the stator. The stator redirects fluid to the pump and increases torque.

UNIT 4
Torque Converters

True-False Questions (Indicate the correct choice)

1. The torque converter can multiply torque. T F

2. Torque is a word used to describe a turning or twisting effort applied to something to make it turn or try to make it turn. T F

3. The shorter the wrench, the greater the force or torque that can be applied to a bolt. T F

4. Torque can be multiplied through gears. T F

5. When torque is multiplied, speed is also increased. T F

6. The stator in a torque converter redirects fluid back toward the pump to help it turn. T F

7. The stator is used only at low vortex. T F

8. The torque converter multiplies engine torque in proportion to the difference in speed between the pump and turbine. T F

9. Torque is multiplied in a torque converter because stationary blades on the stator redirect fluid back to the pump to help it turn. T F

10. The torque converter minimizes the necessity of using so many gears to get the vehicle in motion. T F

Multiple Choice Questions (Indicate the correct choice)

1. If 20 pounds of force is applied to a wrench that is one foot long, the torque exerted would be:
   a. 1 pound-foot c. 20 pounds-feet
   b. 10 pounds-feet d. 40 pounds-feet

2. If 20 pounds of force is applied to a wrench that is two feet long, the torque exerted would be:
   a. 1 pound-foot c. 20 pounds-feet
   b. 10 pounds-feet d. 40 pounds-feet

3. Which of the following is *not* one of the components of a torque converter?
   a. pump c. reactor
   b. turbine d. stator

4. The torque converter component that is attached to the engine is the:
   a. pump b. turbine c. stator

5. The torque converter component that is attached to the input shaft of the planetary gearbox is:
   a. pump b. turbine c. stator

6. The stator clutch is locked to allow the stator to redirect fluid to the pump during:
   a. low vortex b. high vortex c. zero vortex

Parts Identification

Identify the components of a torque converter by writing their names in the spaces provided.

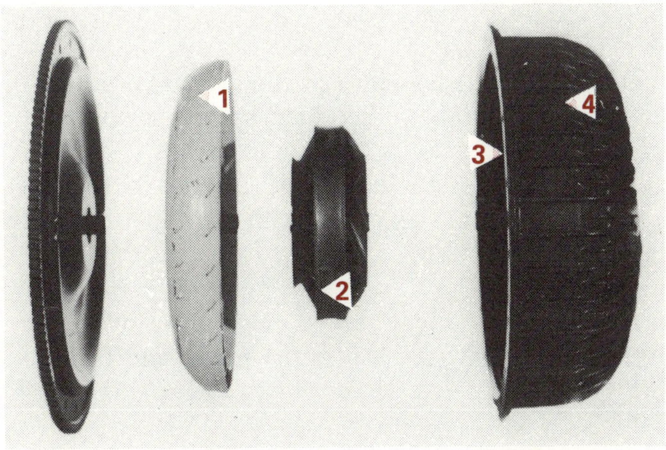

1.________________ 2.________________ 3.________________ 4.________________

UNIT 4
Torque Converters

True-False Questions

1. This statement is true. The torque converter has all the advantages of a fluid coupling, and, in addition, it can increase or multiply torque.

2. This statement is true. When a mechanic uses a wrench to tighten a nut or bolt he is applying a torque on the bolt or nut through the wrench.

3. This statement is false. The shorter the wrench, the lesser the force or torque that can be applied to a bolt. To exert more torque, a longer wrench must be used.

4. This statement is true. Just as torque can be multiplied with a longer wrench, it can also be multiplied with larger gears.

5. This statement is false. Torque is gained at the expense of a loss in rpm. Torque increase means speed reduction, rather than speed increase.

6. This statement is true. The stator has blades curved in such a way that they change the direction of the fluid as it leaves the turbine and redirect it toward the pump in a helping direction.

7. This statement is false. The stator is needed only during time of *high* vortex. The stator is not needed at low vortex because the pump and turbine are turning at nearly the same speed, and there is very little vortex flow to be redirected.

8. This statement is true. When vortex is high there is a good deal of torque multiplication. When vortex is low there is very little torque multiplication.

9. This statement is true. Torque multiplication—or torque conversion—is possible because the stator redirects fluid to the pump and increases torque.

10. This statement is true. Engine torque which enters the converter can be increased as it goes out. The amount of gear reduction in the planetary gearbox can be minimized.

Multiple Choice Questions

1. The correct choice is *c*. Torque is calculated by multiplying the pounds of force times the length of the lever (20 × 1 = 20 pounds-feet).
2. The correct choice is *d*. The 20 pounds of force multiplied by the two-foot wrench length equals 40 pounds-feet.
3. The correct choice is *c*. The main components of a torque converter are the pump, turbine, and stator.

4. The correct choice is *a.* The pump is attached to the engine.

5. The correct choice is *b.* The turbine is attached to the input shaft of the planetary gearbox.

6. The correct choice is *b.* The stator is locked or unlocked according to vortex conditions. During conditions of high vortex, the fluid leaves the turbine at an angle to strike the stator squarely on the front of its blades. Pushing the stator in this direction locks up the one-way clutch, and fluid is redirected to the pump.

Parts Identification

The names of the torque converter components are:

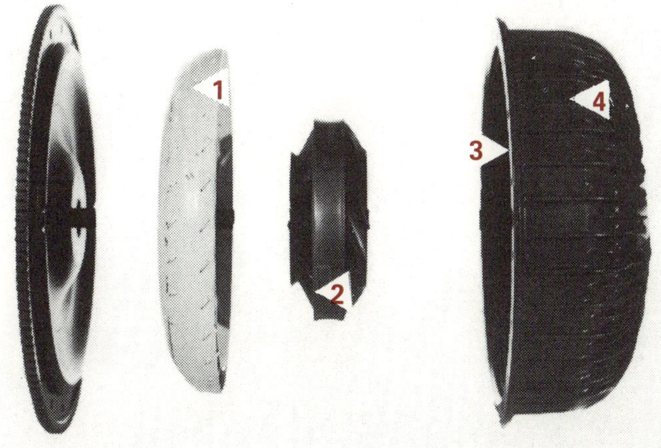

1. turbine 2. stator 3. pump 4. housing

Section III

PLANETARY GEARBOX

Planetary Systems for Automatic Transmissions

Automatic transmissions use a type of gear arrangement called a planetary gear system. Planetary gearing was not a new idea invented for automatic transmissions, but was used in various devices long before the advent of the automatic transmission. Each of the 15 million model "T" Fords, in fact, used a planetary system. Early vehicles had certain problems with this type of gearing. There were difficulties in designing systems which would provide enough forward gear reductions, and it was hard to get the gearbox to operate without excessive chattering. As a result, the planetary transmission faded from the automotive scene until the late 1930s, when it reappeared in the automatic transmission.

Figure 5-1

Components of a Simple Planetary Gear Set

Simple Planetary Systems

There are two types of planetary systems—simple planetary systems and compound (or complex) planetary systems. The simple planetary system was the type used in early vehicles. It has been used in overdrive units attached to the rear of standard transmissions, and is the kind of planetary system used on many automatic transmissions. An understanding of the simple planetary gear system is basic to an understanding of more complex systems.

A simple planetary gear system is constructed from four components: (1) a sun gear, (2) a set of planets, (3) a carrier, and (4) a ring gear. The relationship of the components is shown in figure 5-1, and a brief description of their functions follows.

The Sun Gear. The sun gear gets its name from the central position it

Figure 5-2 Sun Gear

occupies in a planetary gear set. In a simple planetary gear set the sun gear may have one of three functions: (1) it may be the gear to which input force is applied (the engine, for example, may be connected to it), in which case it is called the *turning* gear; (2) it may be the output or *driven* gear; or (3) it may be *held stationary*.

The Planets and Planet Carrier. The planets get their name from the fact that they revolve around the sun gear. The carrier supports the planets. The planets

Figure 5-3 **Planets and Planet Carrier**

(sometimes called pinions) are in mesh with both the sun gear and the ring gear. The planets are mounted in the carrier in such a way that they may "walk" around the sun gear while the carrier rotates, or they may "idle" on the stationary carrier while the sun and ring gears rotate. Like the sun gear, the carrier assembly may perform one of three functions: (1) it may be used as the input or *turning* member, (2) it may serve as the output or *driven* member, or (3) it may be *held stationary*.

The Ring Gear. The ring gear has internal teeth which mesh with the planets. Some manufacturers call this gear an *internal* gear or an *annular* gear.

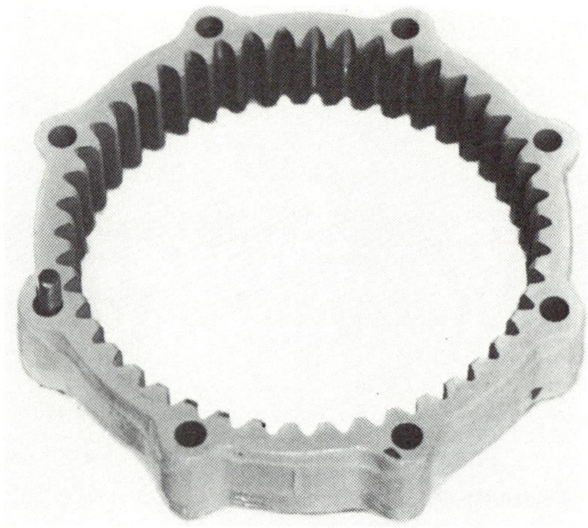

Figure 5-4 **Ring Gear**

Like the sun gear and planet gears, the ring gear may perform one of three functions: (1) it may act as the input or *turning* gear, (2) it may serve as the output or *driven* gear, or (3) it may be *held stationary*.

Planetary Gear Operation

It is possible to arrive at six different conditions of gear reduction or overdrive in a simple planetary gear set. To do this, one need only vary the gear which is turned, driven, or held stationary. An explanation of each of these conditions follows:

Condition One. The first condition is to turn the planet carrier and hold the sun gear stationary. The ring gear becomes the output or driven member. Figure 5-5 shows what happens under

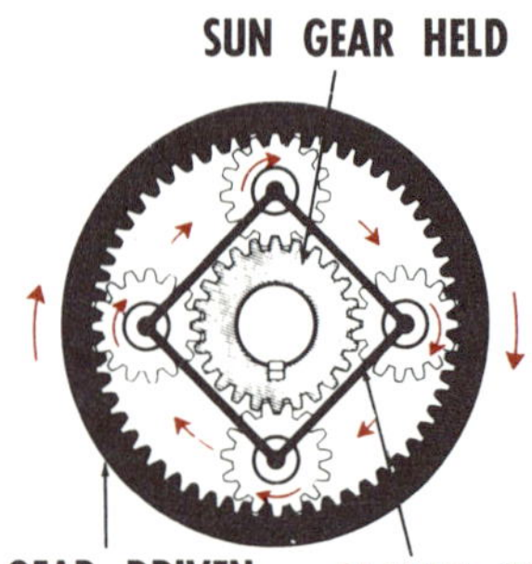

Figure 5-5

***Condition One:* Forward Increase**

these conditions. The planets walk around the stationary sun, and drive the ring gear. The ring gear is driven at a greater speed than the input to the carrier. This is an overdrive condition which would not be applicable to an automatic transmission because a gear reduction is required to multiply torque.

Condition Two. A second possibility is to turn the ring gear and hold the sun gear stationary. Now the planet-carrier assembly becomes the output or driven member. As figure 5-6 demonstrates, the

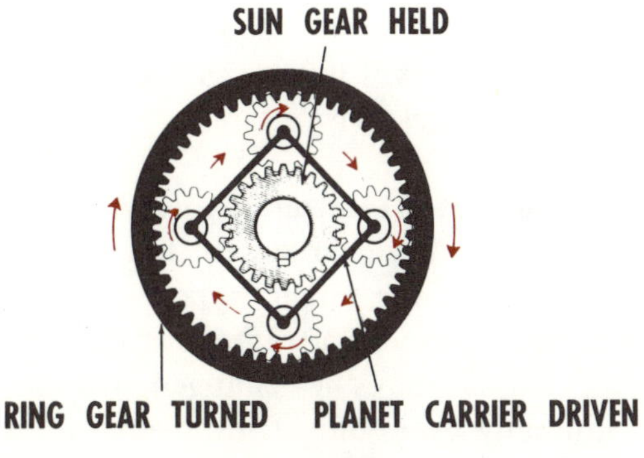

Figure 5-6

Condition Two: **Forward Reduction**

ring gear forces the planets to walk around the stationary sun, thereby rotating the carrier. The carrier turns more slowly than the ring gear, creating a condition of gear reduction and torque multiplication. This condition is one that could be used in an automatic transmission.

Condition Three. Condition three requires that the sun gear be the input, or turning, member, while the ring gear be held stationary. In this case, the output, or driven, gear is the carrier. Figure 5-7 shows that as the sun gear rotates, it forces the planets to walk around the stationary ring gear, driving the carrier in reduction. Since this condition causes a gear reduction and torque multiplication, it is applicable to an automatic transmission.

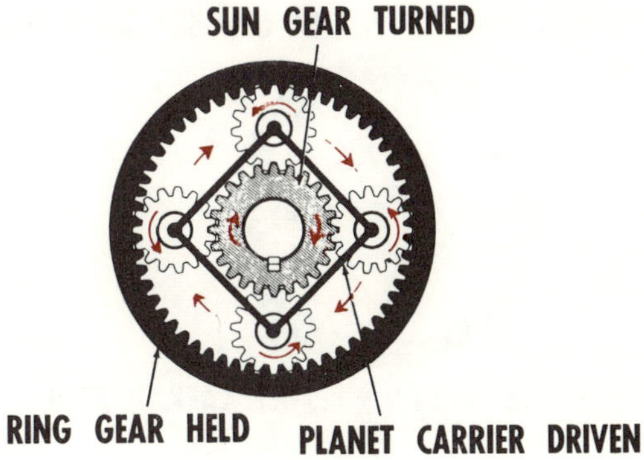

Figure 5-7

Condition Three: **Forward Reduction**

Condition Four. In condition four, the driving or input member is the planet carrier. The ring gear is held stationary, and the output or driven member is the sun gear. As figure 5-8 demonstrates, the

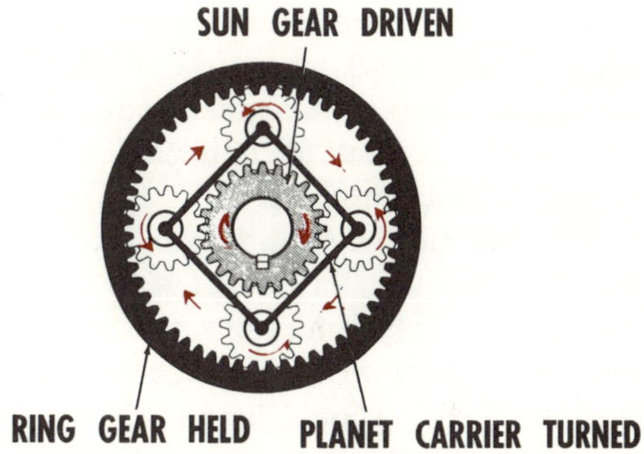

Figure 5-8

Condition Four: **Forward Increase**

planets walk around the ring gear and drive the sun gear at an increase in speed. This is an overdrive condition and is not appropriate for an automatic transmission.

Condition Five. In condition five, the stationary member is the planet carrier. The turned or input member is the ring gear. The output is taken from the sun gear. In a simple planetary gear set, holding the planet carrier causes the planets to act as idlers, creating a change

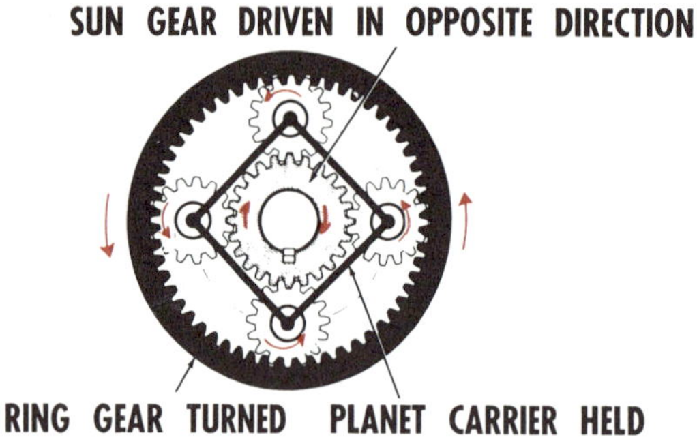

Figure 5-9

Condition Five: Reverse Increase

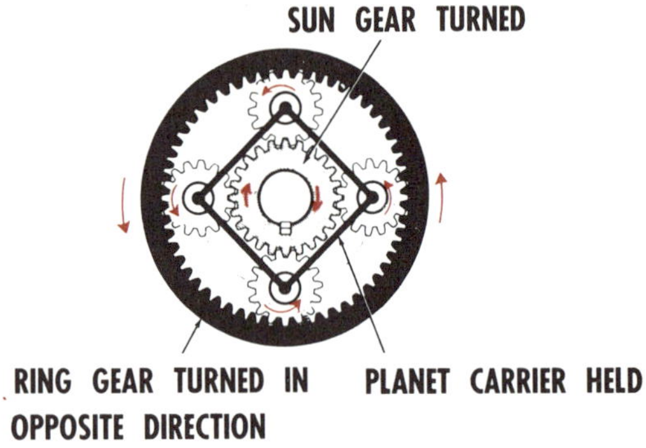

Figure 5-10

Condition Six: Reverse Reduction

in direction. As figure 5-9 shows, the ring gear drives the sun gear through the planets in a reverse direction. It is a reverse overdrive, which would not be useful in an automatic transmission.

Condition Six. Condition six again calls for the carrier to be held stationary. This time, however, as figure 5-10 shows, the sun gear is turned and the ring gear is driven. This causes a reversing of direction again, but with a gear reduction. This condition would be applicable to an automatic transmission.

The six conditions can be summarized by the Summary Table, figure 5-11. Notice that only conditions two, three, and six are appropriate for an automatic transmission. It should be added that if no member of the planetary gear system is held, the result is that no power flows through the system. This is how neutral is achieved in an automatic transmission. When any two of the major planetary components are caused to rotate in the same direction at the same speed, the third member will follow, causing a condition of direct drive.

Figure 5-11

Summary Table

CONDITIONS	1	2	3	4	5	6
RING GEAR	D	T	H	H	T	D
CARRIER	T	D	D	T	H	H
SUN GEAR	H	H	T	D	D	T
SPEED	I	L	L	I	IR	LR

Advantages of a Planetary Gear System

The planetary gear system has several advantages over other types of gear systems. The main advantage is that the gear ratios may be changed without taking any gears out of mesh—no gears have to be moved in or out of place. As explained previously, the six different conditions are achieved by holding and turning different members of the planetary system. Another important advantage of planetary gears is that they are small and compact. This means that the gearbox section of the automatic transmission will be smaller and lighter than it would be if another gear system was used. Finally, the planetary gear set is strong because the load can be distributed over several gears.

Complex Planetary Systems

Simple planetary gear systems work satisfactorily, but they occupy a large amount of space. The planetary gearbox had to be quite large and, as a result, was quite heavy. The replacement of the fluid coupling with a torque converter, which could help in the job of torque multiplication, made fewer gear reductions feasible. Four-speed automatic transmissions began to disappear in favor of three-speed and even two-speed systems. Engineers began to look at ways of making the gearbox smaller, and the result was the complex planetary gear system.

A complex planetary gear set combines two or more simple planetary gear sets into one unit. The complex planetary set works just like the simple planetary set, but it does so with more flexibility, and it occupies less space as well. A typical complex planetary system is shown in figure 5-12.

This complex planetary unit uses one sun gear, called a compound sun gear. Two sets of planets are in mesh with the sun gear, and a ring gear is in mesh with each of the planet sets. Which of the units is held or driven depends on what the manufacturer wishes to accomplish with the gear train.

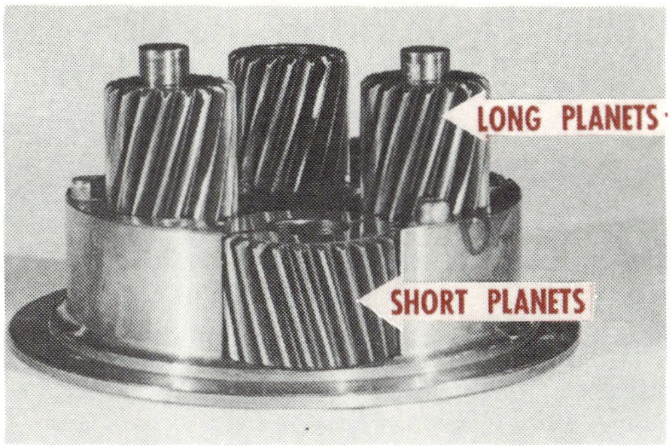

Figure 5-13

Complex Planetary with Long and Short Planets

Another complex planetary gear arrangement is shown in figure 5-13. This unit also uses a compound sun gear. In mesh with the compound sun gear is a set of long and short planets. Each set is in mesh with one of the sun gears. One ring gear and one carrier are used. This unit also may be used in a variety of ways depending on the manufacturer's requirements.

Figure 5-12

Complex Planetary Gear Set with Two Sets of Planets

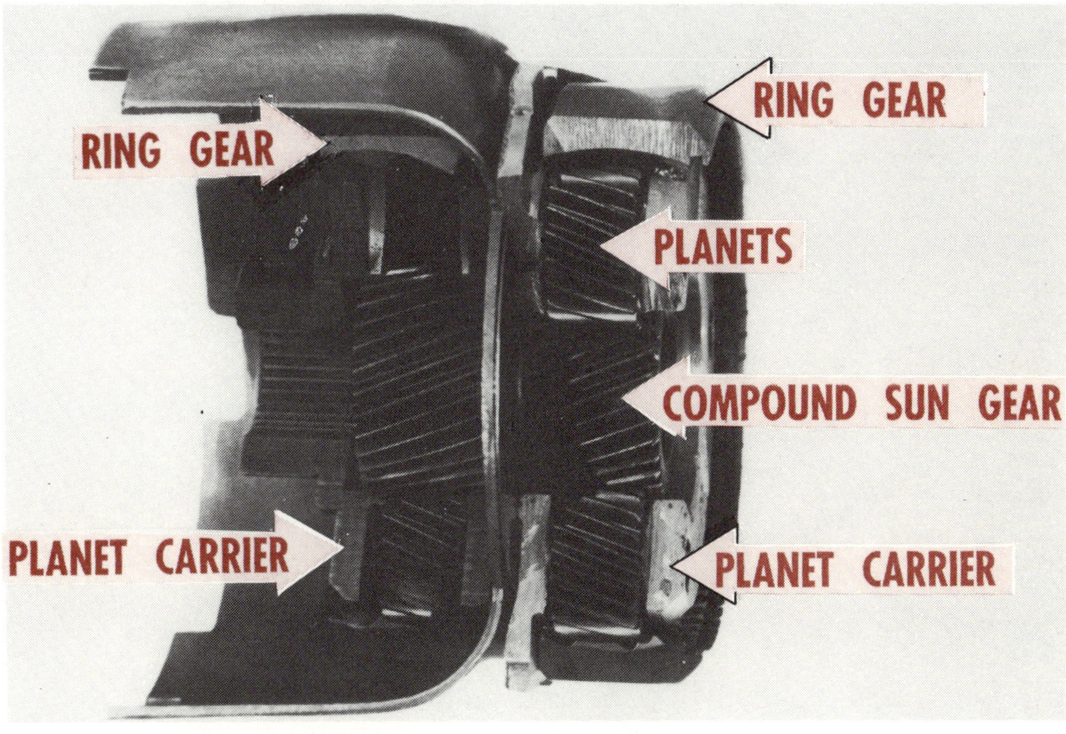

Application

The complex planetary gear set is now used in all late-model automatic transmissions. The transmission needs only one complex planetary gear set to get all the forward and reverse reductions required.

The complex planetary gear set offers the same advantages as the simple planetary—shifting without taking any gears out of mesh, and distribution of the load over several gears; but the complex planetary allows for more flexibility in design.

UNIT 5
Planetary Systems for Automatic Transmissions

True-False Questions (Indicate the correct choice)

1. Planetary gear sets are a relatively new innovation (since 1940). T F

2. In a planetary gear set, gear ratio changes are accomplished without taking any gears out of mesh. T F

3. In a simple planetary gear set, the number of possible forward gear reductions is two. T F

4. Neutral is accomplished in a planetary gear set by holding only one unit. T F

5. When making the gear changes in a simple planetary set, all holding is done on the ring gear or planet carrier. T F

6. Holding any two units in a planetary gear set will result in a reverse. T F

Multiple Choice Questions (Indicate the correct choice)

1. When any two members of a planetary gear set are locked together, the ratio would be:
   a. 3 to 1
   b. 2 to 1
   c. 1 to 1
   d. none of the above

2. A simple planetary gear set can be operated by:
   a. changing power input to a different member
   b. obtaining the power output from a different member
   c. holding one of the members
   d. all of the above

3. In planetary gear systems, the main principle involved in all combinations where there is *gear reduction* is:
   a. Drive one member and hold the other two.
   b. Drive one member and hook one of the other two to the output.
   c. Drive one member, take the output torque off another, and hold the third member.
   d. Drive either the sun gear or the ring gear, and always hold the planet carrier.

4. In a simple planetary gear set, if the ring gear is driven, the carrier turned, and sun gear held, the result is:
   a. forward increase
   b. forward reduction
   c. reverse increase
   d. reverse reduction

5. In a simple planetary gear set, if the ring gear is turned, carrier held, and sun gear driven, the result is:
   a. forward increase
   b. forward reduction
   c. reverse increase
   d. reverse reduction

6. In a simple planetary gear set, if the ring gear is turned, carrier driven, and sun gear held, the result is:
   a. forward increase
   b. forward reduction
   c. reverse increase
   d. reverse reduction

7. Which of the following is an advantage of a planetary gear system?
   a. small and compact
   b. in constant mesh
   c. load distributed over several gears
   d. all of the above

Planetary Gearing Problems

Indicate on the drawing by arrows which direction each of the members will revolve. Write in the space given what the effect will be on the output gear. (See example number 1.)

1. Stationary Member—Ring Gear
   Turning Member—Sun Gear
   Driven Member—Planet Carrier
   Effect: Reduction

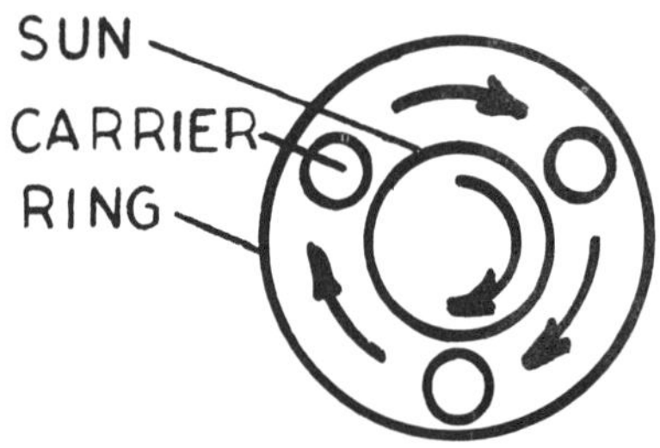

2. Stationary Member—Ring Gear
   Turning Member—Planet Carrier
   Driven Member—Sun Gear
   Effect: ____________________

3. Stationary Member—Sun Gear
   Turning Member—Ring Gear
   Driven Member—Planet Carrier
   Effect: ____________________

4. Stationary Member—Sun Gear
   Turning Member—Planet Carrier
   Driven Member—Ring Gear
   Effect: ____________________

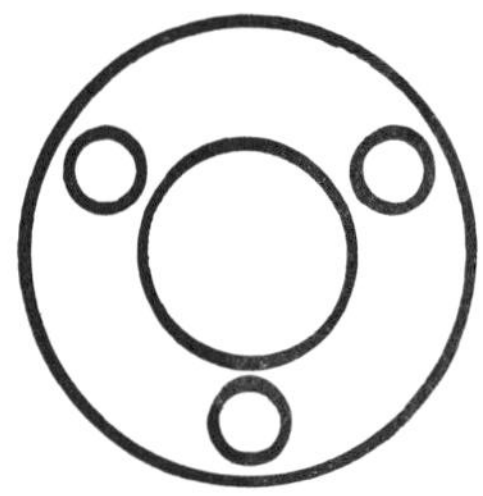

5. Stationary Member—Planet Carrier
   Turning Member—Sun Gear
   Driven Member—Ring Gear
   Effect: ____________________

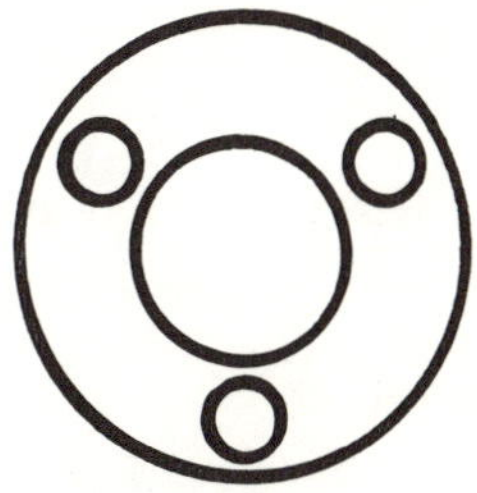

6. Stationary Member—Planet Carrier
   Turning Member—Ring Gear
   Driven Member—Sun Gear
   Effect: ____________________

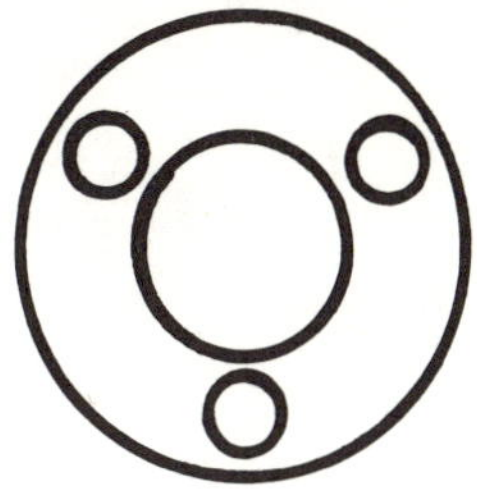

Parts Identification

1. Identify the major components of a simple planetary gear set.

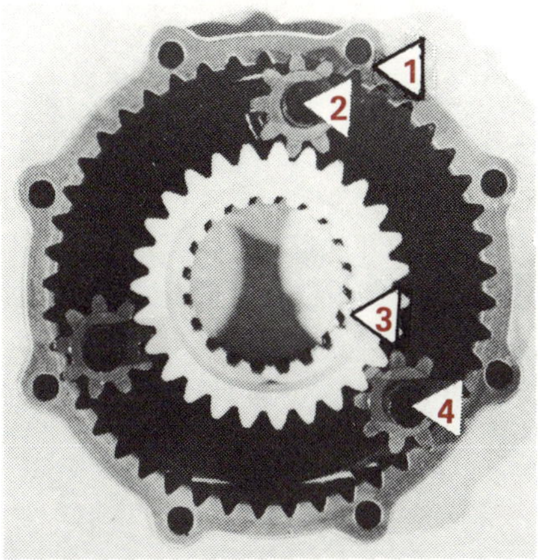

1. ____________________________
2. ____________________________
3. ____________________________
4. ____________________________

2. Identify the components of a complex planetary gear set.

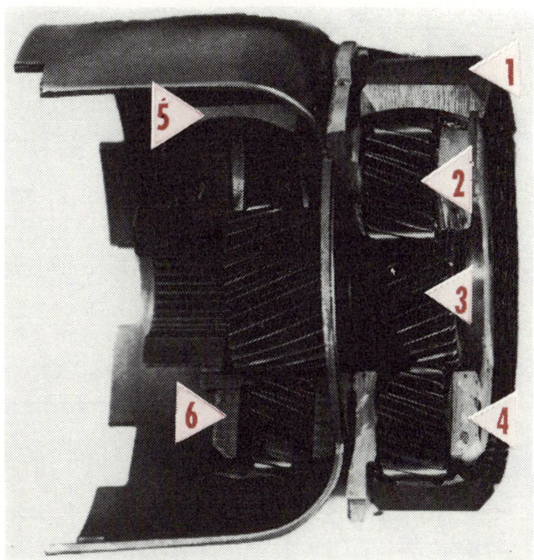

1. ____________________________
2. ____________________________
3. ____________________________
4. ____________________________
5. ____________________________
6. ____________________________

UNIT 5
Planetary Systems for Automatic Transmissions

True-False Questions

1. This statement is false. Model "T" Fords were equipped with planetary gear sets, so they are not a relatively new innovation.

2. This statement is true. One of the advantages of a planetary gear set is that gear changes can be accomplished without taking any gears out of mesh. All the gears in a planetary gear set are constantly in mesh.

3. This statement is true. There are only two forward gear reductions possible in a simple planetary gear set.

4. This statement is false. Neutral is accomplished in a planetary gear set by not holding any of the units.

5. This statement is false. Any of the three members can be held in a simple planetary gear set to make gear changes. All the holding does *not* have to be done on the ring gear or planet carrier.

6. This statement is false. Holding any two units in a planetary set would lock it up so it would rotate as a unit. Reverse is only possible by holding the carrier unit.

Multiple Choice Questions

1. The correct choice is *c*. When two units of a planetary gear set are locked together the gear system rotates as a unit with no gear reduction.

2. The correct choice is *d*. A simple planetary gear set can be operated by doing any one of the following: changing power input to a different member, obtaining the power output from a different member, or holding one of the members.

3. The correct choice is *c*. Any time there is a gear reduction one member must be driven, the output taken from another, and a third member held.

4. The correct choice is *a*. In a simple planetary gear set, if the ring gear is driven, the carrier turned, and the sun gear held, the result is a forward increase.

5. The correct choice is *c.* In a simple planetary gear set, if the ring gear is turned, carrier held, and sun gear driven, the result is a reverse increase.

6. The correct choice is *b.* In a simple planetary gear set, if the ring gear is turned, carrier driven, and sun gear held, the result is a forward reduction.

7. The correct choice is *d.* The advantages of a planetary gear set are: (1) the unit it small and compact; (2) the unit is in constant mesh; and (3) the load is distributed over several gears.

Planetary Gearing Problems

2. Stationary Member—Ring Gear
   Turning Member—Planet Carrier
   Driven Member—Sun Gear
   Effect: increase

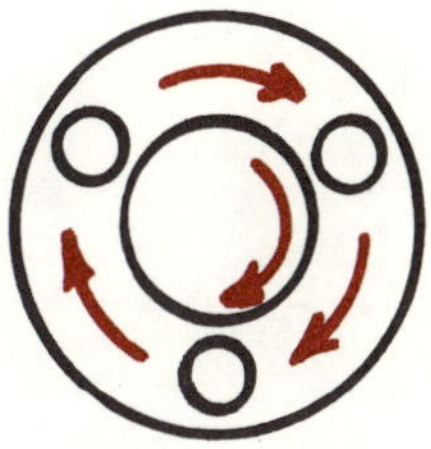

3. Stationary Member—Sun Gear
   Turning Member—Ring Gear
   Driven Member—Planet Carrier
   Effect: reduction

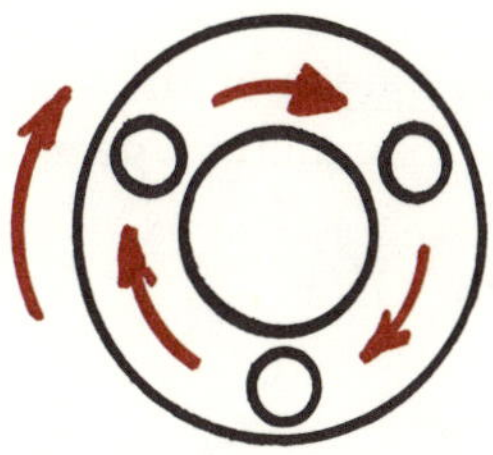

4. Stationary Member—Sun Gear
   Turning Member—Planet Carrier
   Driven Member—Ring Gear
   Effect: increase

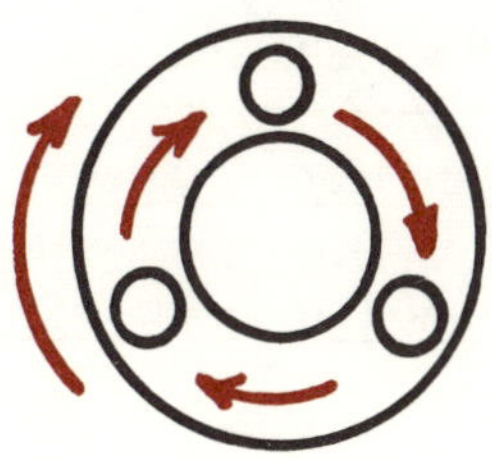

5. Stationary Member—Planet Carrier
Turning Member—Sun Gear
Driven Member—Ring Gear
Effect: reverse reduction

6. Stationary Member—Planet Carrier
Turning Member—Ring Gear
Driven Member—Sun Gear
Effect: reverse increase

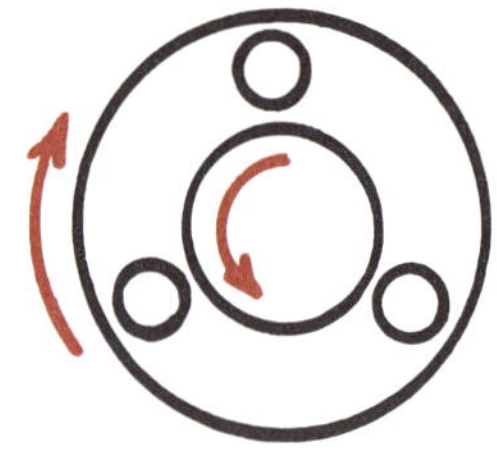

Parts Identification

1. The components of a simple planetary gear set are:

1. Ring Gear
2. Planet Carrier
3. Sun Gear
4. Planet

2. The correct names of the components of the complex planetary gear set are:

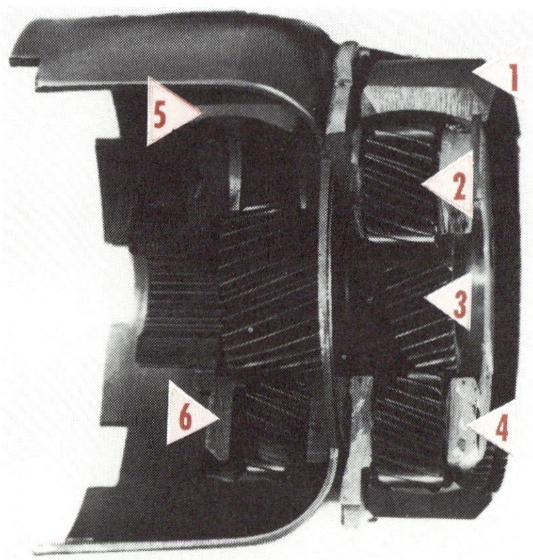

1. Ring Gear
2. Planets
3. Compound Sun Gear
4. Planet Carrier
5. Ring Gear
6. Planet Carrier

Planetary Holding Units

Gear ratio changes in a planetary gearbox are possible because various members of the planetary set are held from turning. The devices which do this holding in most automatic transmissions are multiple disc clutches, bands and servos, and one-way clutches. The operation of these devices is presented in this unit.

Multiple Disc Clutches

Most automatic transmissions use one or more multiple disc clutches to hold planetary gear members and to connect and disconnect various members in the gear train. A multiple disc clutch is composed of a piston, a return spring or springs, a set of friction discs, a mating set of steel discs, and a clutch drum. An exploded view of a typical clutch assembly—or clutch pack—is shown in figure 6-1.

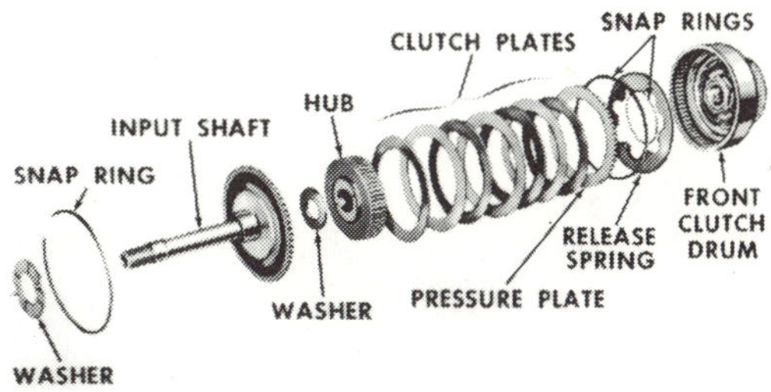

Figure 6-1

Exploded View of a Multiple Disc Clutch
(Ford Motor Co.)

The multiple disc clutch operates to lock or unlock the drum (and the planetary member connected to it) with another planetary member. Figure 6-2

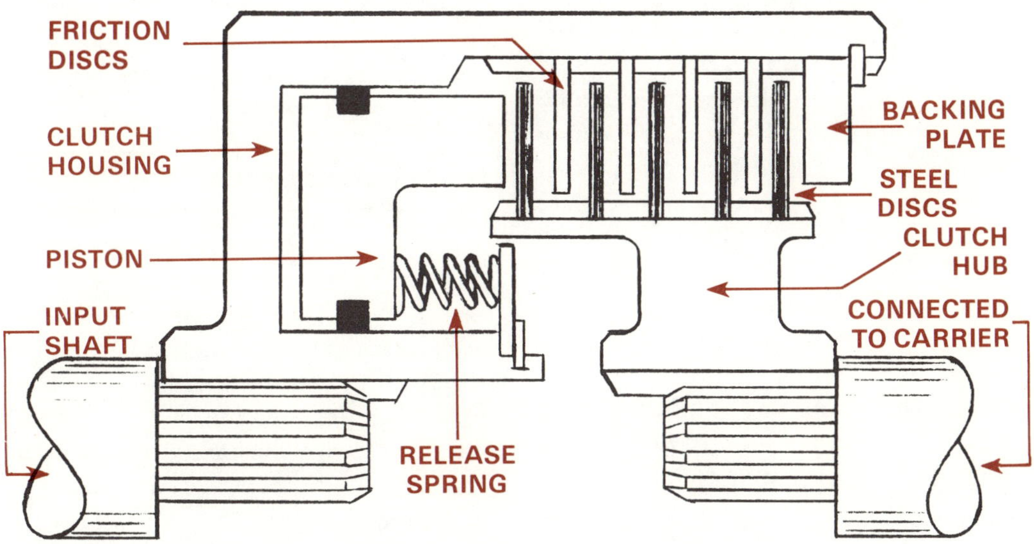

Figure 6-2

Operation of a Multiple Disc Clutch

shows a typical application of a multiple disc clutch. In this case, a planetary carrier is the member to be locked or unlocked with the clutch drum. The steel discs are splined to the drum and the friction discs are splined to the carrier. When hydraulic pressure is applied to the piston, it squeezes all the discs together. As a result, the drum is locked to the carrier. When hydraulic pressure is removed from the piston, the return spring moves it back. The clutch discs separate, unlocking the drum from the carrier.

The primary advantage of a multiple disc clutch is that it allows for a great deal of frictional area in a small package. Other advantages include: (1) a flexibility made possible by a wide range of frictional materials; (2) no necessity to adjust for wear; (3) the possibility of modifying the capacity of the clutch by adding or removing clutch plates; and (4) unlike a band, highly stressed areas in the transmission case are eliminated.

Two basic types of frictional materials are used on the surfaces of the friction plates. These are sintered metallic materials and non-metallic materials. The metallic plate consists of metallic powders sintered to the plate. The heat resistance and durability of the metallic plates are excellent. The sintered facing is porous, permitting absorption of fluid. The disadvantages of this kind of plate are the cost—due to the high cost of the copper powder used—and the lack of compressibility. Because this plate cannot be compressed easily, high spot contact is made on the mating steel plates. This causes a rather erratic frictional contact not present with a more compressible facing material.

The non-metallic materials used for clutch facings are cork, cotton, asbestos, and paper. Some paper materials have higher coefficients of friction than the sintered metallic plate. The cost of paper plates is lower, and the contact pattern is excellent because of the compressibility of paper. Most late-model automatic transmissions use a non-metallic friction material. An assortment of clutch discs is pictured in figure 6-3. The grooving

Figure 6-3

Sample Clutch Plates or Discs

pattern on the faces of the clutch plates is designed to aid in cooling. The clutch is run wet in automatic transmission fluid, and the fluid is allowed to circulate in the grooves to carry off the heat.

The apply piston for the clutch pack is most commonly machined from aluminum. Since hydraulic pressure must be contained behind the piston, some type of seal must be used around the piston to seal it in its bore. Several types of seals are in use. The three main seal types are shown in figure 6-4. The iron ring is the oldest type of seal. Its primary advantages are durability and temperature resistance. However, its leakage is usually greater than that of seals made of synthetic compounds. Late-model transmissions use a synthetic lip or "O" ring seal. Although the synthetic seals do a

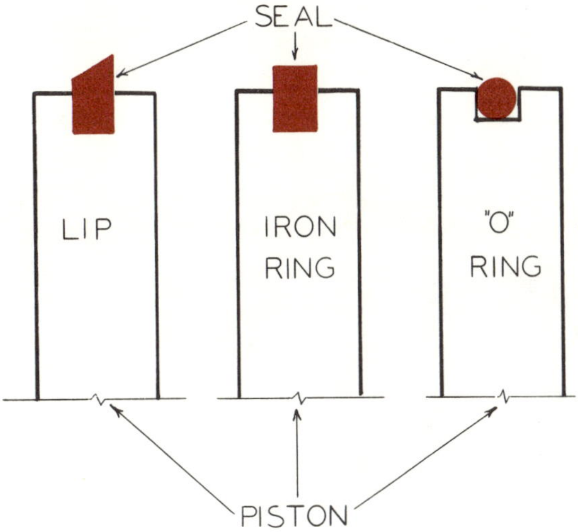

Figure 6-4

Types of Piston Seals

good job of preventing leakage, they are more subject to wear than iron seals.

The types of return springs which force the piston to the release position are shown in figure 6-5. The clutch piston may be returned by means of a large coil spring, a series of small coil springs, a Belleville spring, or a waved washer.

Figure 6-5

Types of Clutch Return Springs

Band and Servo Assembly

Bands are another important holding device used in the planetary gearbox. Most automatic transmissions use one or two bands. A band and servo assembly is shown in figure 6-6. The unit consists of a

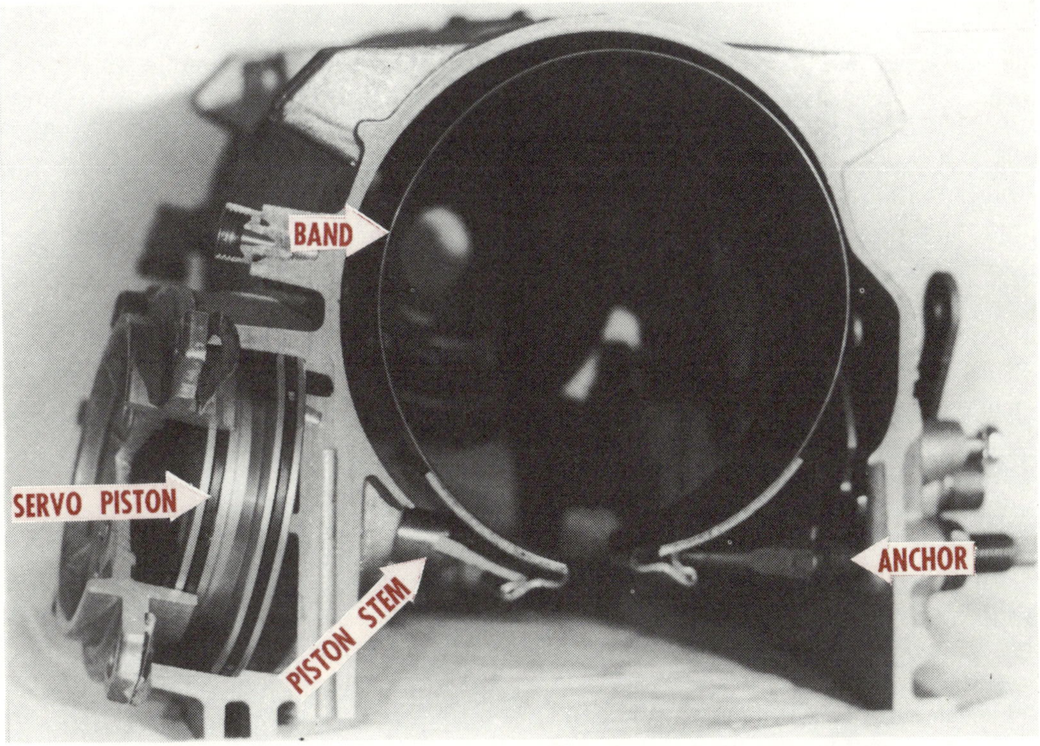

Figure 6-6

Band and Servo Assembly

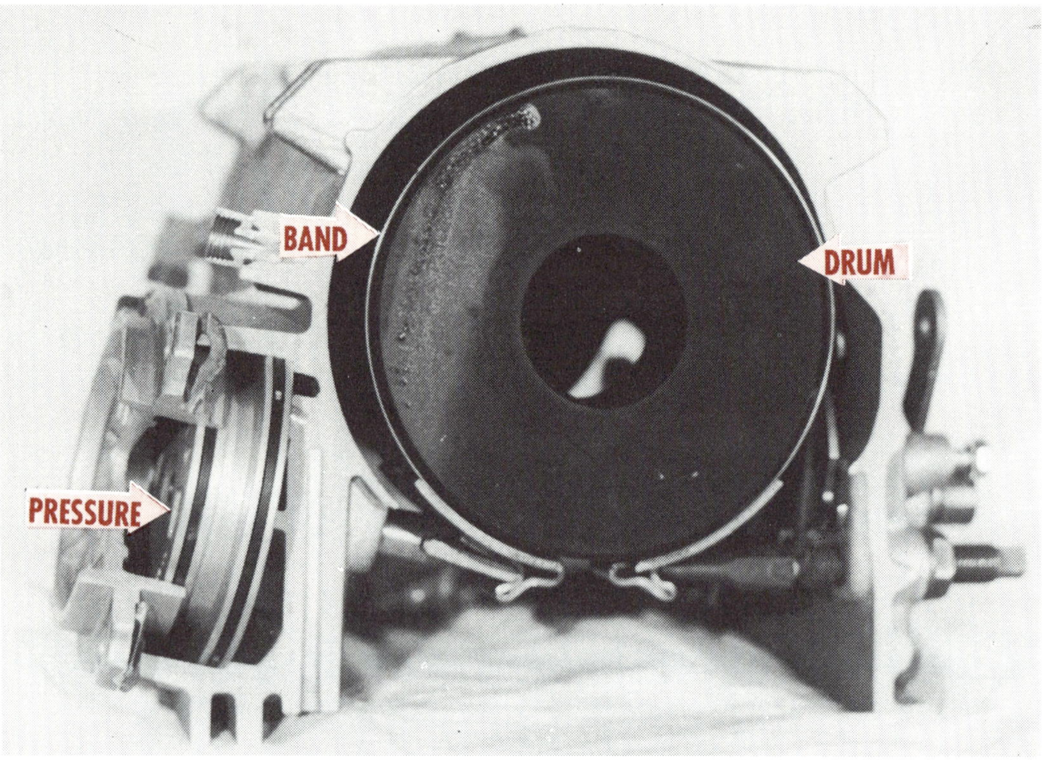

Figure 6-7

Band Operation

band, a servo piston, a piston stem, an anchor, and a piston return spring.

The band operates to stop rotation of a planetary member. When the band is activated or applied (see figure 6-7), fluid under pressure is routed to the servo piston. Pressure on the servo piston moves it in its bore. The piston stem attached to the piston also moves and pushes on the band. Since the band is anchored at the other end, this action tightens the band around the drum which is attached to a planetary member. This action stops the rotation of the drum. When the band is disengaged, pressure is removed from the piston, and the return spring pushes the piston back to relieve the tension on the band and free the drum.

Like the multiple disc clutch, the band and servo assembly offer the advantage of a good deal of frictional contact in a small space. Bands in use today are either of the single wrap or double wrap type, shown in figure 6-8. A double wrap band offers the advantages of better holding capacity and better conformity with the drum, and it requires approximately one-quarter the force to create the same

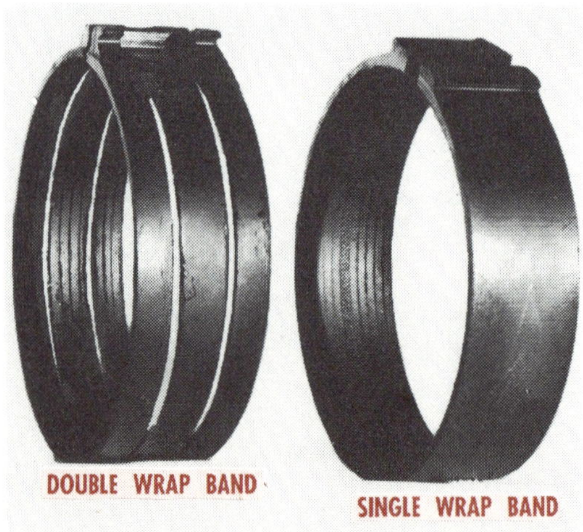

Figure 6-8

Single and Double Wrap Band

holding effort as a single wrap band. Therefore, the servo piston area can be reduced accordingly. The one disadvantage of the double wrap band is its higher cost.

The frictional materials used as a lining on the band are either semi-metallic or organic. Semi-metallic materials are harder and cause considerable wear on the drum. Late-model automatic transmissions use organic frictional materials such as paper and asbestos. These materials cause less wear on the drum, but are subject to more wear themselves. The softer materials offer an additional advantage of conforming better to the drum surface, due to greater flexibility and lack of "high spot" contact possible with semi-metallic band linings.

One-Way Clutches

Most automatic transmissions use one or two one-way clutches in the planetary gearbox to hold planetary members. In addition, the one-way clutch is used in the torque converter to lock and unlock the stator. The one-way clutch is known by various names, including roller clutch, sprag clutch, and overrunning clutch.

The one-way clutch is used to lock up a planet member when it tries to rotate in one direction and allows it to rotate freely in the opposite direction. The one-way clutch consists of an outer race, an inner race, a set of rollers, and a set of springs. The outer race is usually secured to the transmission case, and the inner race connected to a planetary member. As shown in figure 6-9, when the inner race turns clockwise, the rollers are pushed up into the grooves cut into the outer race and the inner race may turn freely. If the inner race were to try to turn counter-clockwise, the rollers would be forced into the narrow part of the grooves and would lock up the inner race.

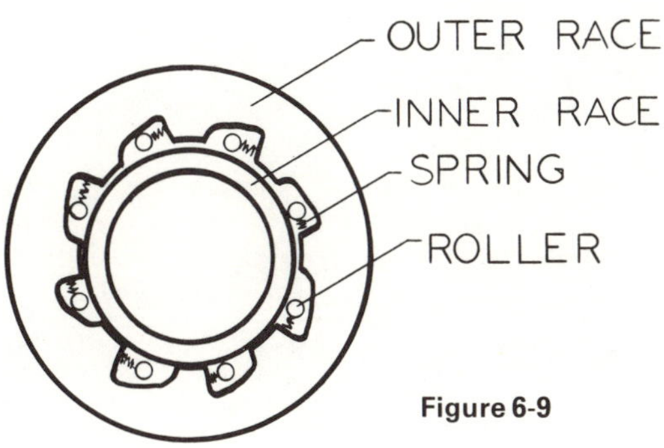

Figure 6-9

One-Way Clutch

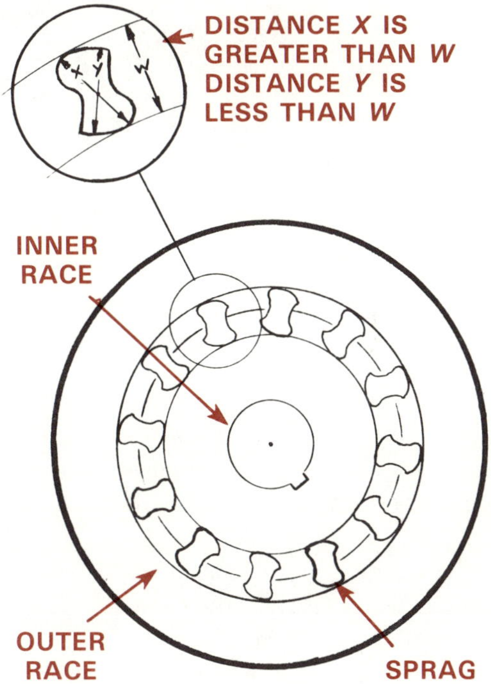

Figure 6-10

Sprag Clutch

Another type of one-way clutch, usually called the sprag clutch, uses sprags instead of rollers. When the outer race attempts to turn to the left with respect to the inner race, the sprags rock up, wedge between the races, and lock up the unit. When the outer race rotates to the right with respect to the inner race, the sprags rock into a position which allows the outer race to rotate. This is illustrated in figure 6-10.

The one-way clutch has several advantages for use in an automatic transmission. First, a one-way clutch provides

Figure 6-11

One-Way Clutch Positioned in Rear of Transmission Case

a good deal of holding force in a small place. Second, a one-way clutch is completely self-contained and there is no need for external controls. Finally, the friction during overrunning is very small so that the clutch does not rob much power from the transmission.

On the other hand, there is a disadvantage with a one-way clutch. Because the device is designed to hold one member from turning relative to another while engine torque is applied, it becomes ineffective when coasting. This is because power is now coming into the transmission from the drive shaft instead of the engine, which causes the clutch to overrun and lose its holding ability. Compression braking (slowing the vehicle by using the engine) is not possible without using some other holding device such as a band or multiple disc clutch. A one-way clutch is pictured in figure 6-11.

UNIT 6
Planetary Holding Units

True-False Questions (Indicate the correct choice)

1. Multiple disc clutches are applied by spring pressure and released by hydraulic pressure. T F

2. The main advantage of a multiple disc clutch is that it allows for a great deal of frictional area in a small package. T F

3. The two basic types of frictional materials used on the surface of the clutch friction plates are metallic and non-metallic. T F

4. Synthetic lip or "O" ring seals are used on clutch pistons because they last longer than iron ring seals. T F

5. The band is applied by hydraulic pressure and released by spring pressure. T F

6. A band and servo assembly offers the advantage of a good deal of frictional contact in a small space. T F

7. The two kinds of bands commonly used are double and single wrap bands. T F

8. The only problem with using a double wrap band is that a greater force is required to apply it than is needed with a single wrap band. T F

9. An advantage of a one-way clutch is that it provides a good deal of holding force in a small space. T F

10. Another advantage of a one-way clutch is that it allows the engine to be slowed by compression braking. T F

Multiple Choice Questions (Indicate the correct choice)

1. The devices which hold the various members of the planetary set from turning to make gear ratio changes are:
   a. multiple disc clutches c. one-way clutches
   b. bands and servos d. all the above

2. The material used for non-metallic friction facing is:
   a. cork b. paper c. asbestos d. all the above

3. The component of a band and servo assembly that is directly acted on by hydraulic pressure is:
   a. anchor   b. band   c. servo piston   d. stem

4. Non-metallic friction materials are most likely to be found on:
   a. band   b. servo   c. clutch piston   d. sprag rollers

5. A planetary holding device which uses rollers and springs is called:
   a. roller clutch   b. one-way clutch   c. overrunning clutch
   d. sprag clutch   e. all the above

Parts Identification

1. Identify the components of a multiple disc clutch.

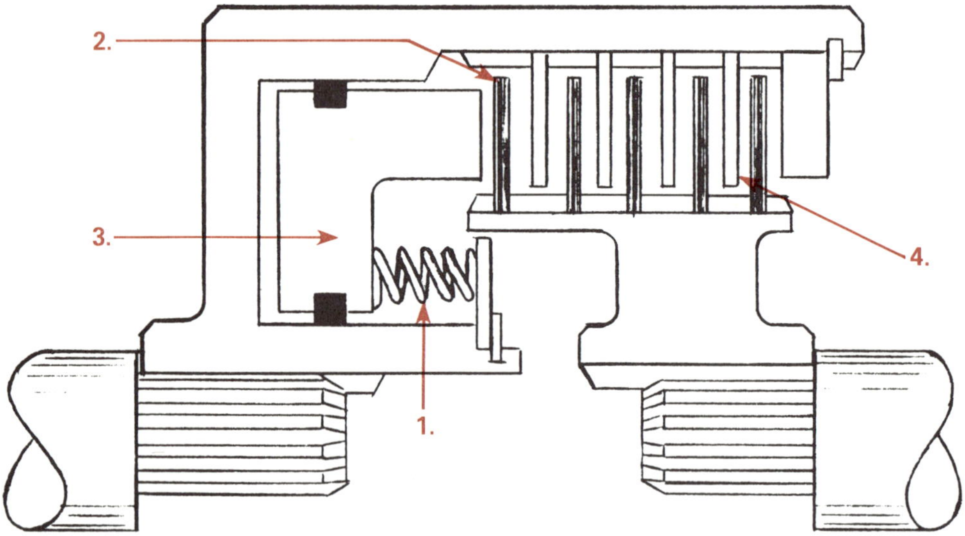

1.________ 2.________ 3.________ 4.________

2. Identify the components of a one-way clutch.

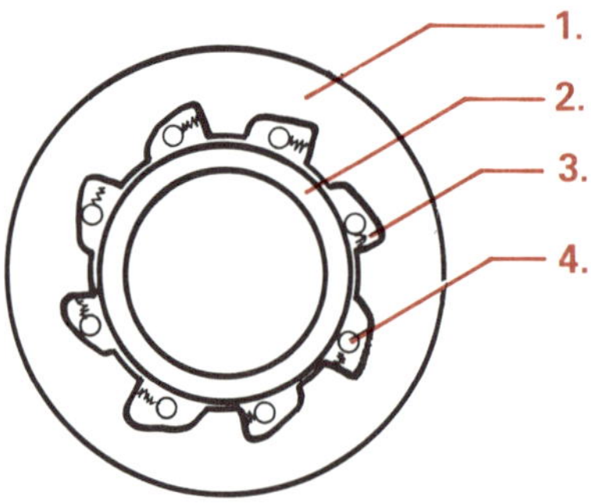

1.________ 2.________ 3.________ 4.________

3. Identify the components of a band and servo assembly.

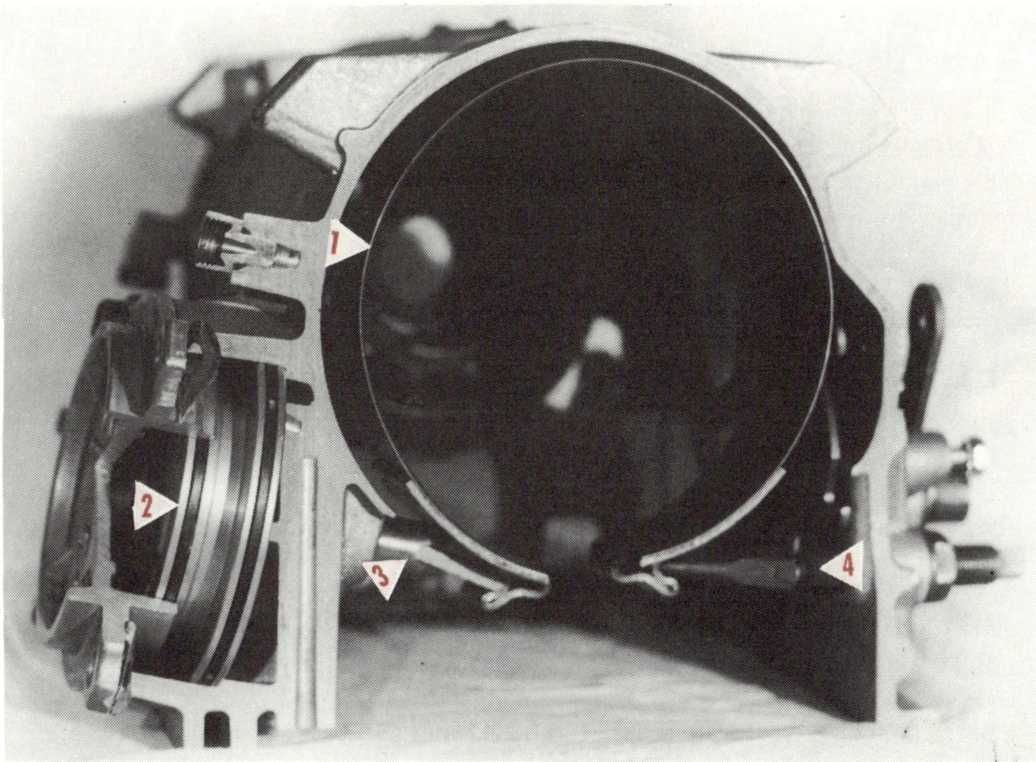

1.______________ 2. ______________ 3. ______________ 4. ______________

UNIT 6
Planetary Holding Units

True-False Questions

1. This statement is false. Hydraulic pressure is used to squeeze the multiple discs together to apply the clutch, and spring pressure is used for release.

2. This statement is true. The multiple disc clutch occupies a small space, but provides a good deal of frictional area.

3. This statement is true. Frictional facings for clutch plates may be made from either metallic or non-metallic materials.

4. This statement is false. The synthetic seals are used on clutch pistons because they do a good job of preventing leakage, but they are more subject to wear than iron seals.

5. This statement is true. Hydraulic pressure acting on one side of the servo piston applies the band, and spring pressure on the other side releases it.

6. This statement is true. Like the multiple disc clutch, the band and servo assembly offers the advantage of a good deal of frictional contact in a small space.

7. This statement is true. Bands in use today are either the single wrap or double wrap type.

8. This statement is false. A double wrap band requires approximately one-quarter the apply force to create the same holding effort as a single wrap band.

9. This statement is true. The one-way clutch occupies relatively little space, but provides a good deal of holding force.

10. This statement is false. The one-way clutch does not allow the engine to be slowed by compression braking. Depending on the power flow, compression braking (slowing the vehicle by using the engine) may not be possible without some other holding device.

Multiple Choice Questions

1. The correct choice is *d.* In most automatic transmissions, holding is accomplished by multiple disc clutches, bands and servos, and sprag clutches.

2. The correct choice is *d.* Cork, paper, and asbestos are non-metallic materials used for multiple disc clutches.

3. The correct choice is *c*. Hydraulic pressure is directed against the servo piston, causing it to move and push on the band.

4. The correct choice is *a*. The frictional material used as a lining on most late-model transmission bands is non-metallic material, such as paper or asbestos.

5. The correct choice is *e*. The one-way clutch used in many automatic transmissions is known by various names, including roller clutch, sprag clutch, and overrunning clutch.

Parts Identification

1. The components of a multiple disc clutch are:

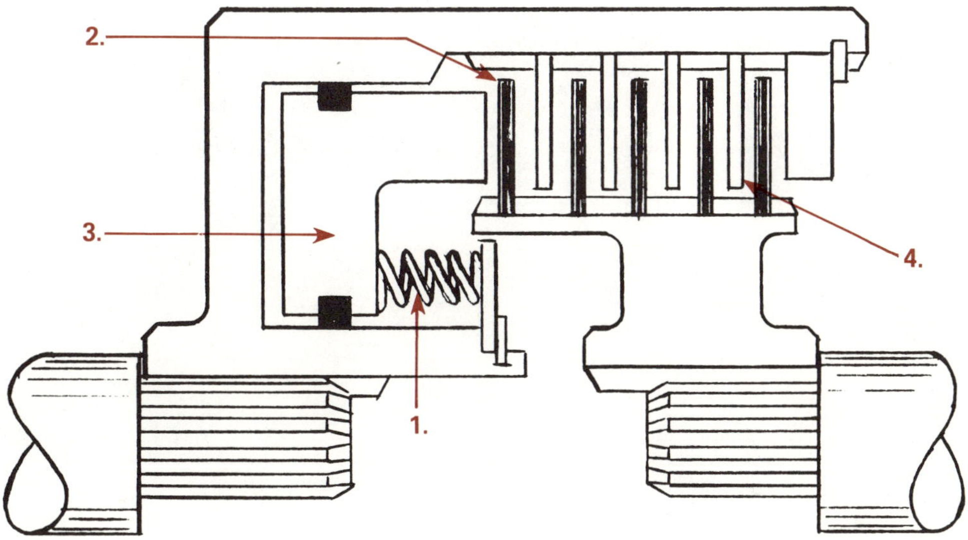

1. Return Spring 2. Friction Disc 3. Piston 4. Steel Disc

2. The components of a one-way clutch are:

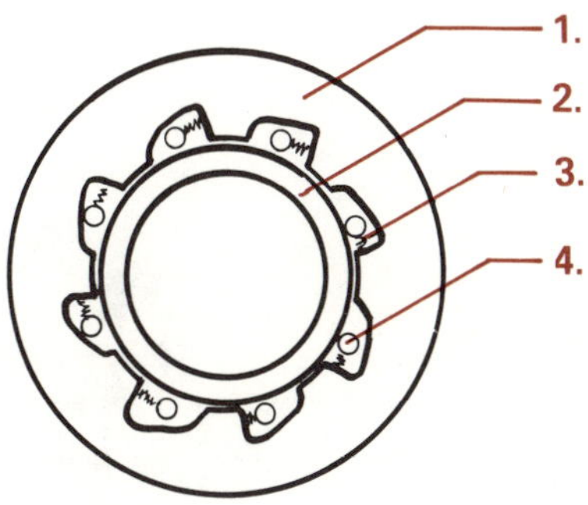

1. Outer Race 2. Inner Race 3. Spring 4. Roller

3 The components of a band and servo assembly are:

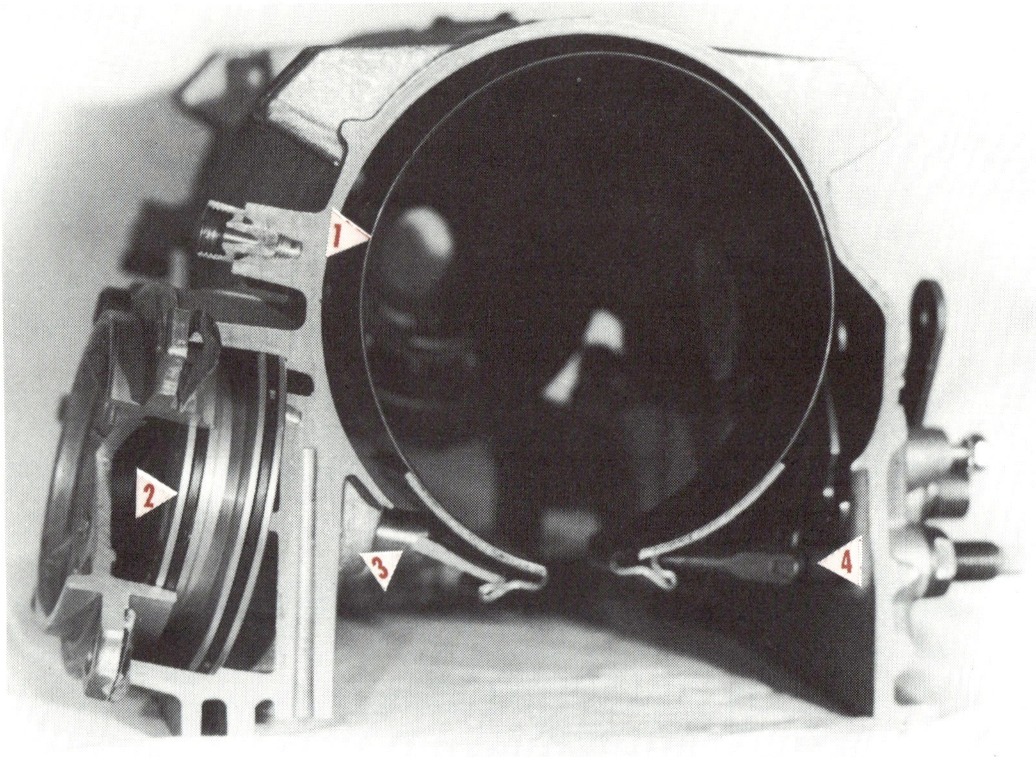

1. Band 2. Servo Piston 3. Piston Stem 4. Anchor

Section IV

THE HYDRAULIC CONTROL SYSTEM

Principles of Hydraulics

The automatic transmission is a hydraulically operated device. It uses the principles of hydraulics to sense and select the correct degree of torque multiplication required by the vehicle. Hydraulics is a very exact and complicated field of study. The science of hydraulics may be divided into two general categories: *hydrodynamics,* the study of moving liquids; and *hydrostatics,* the study of liquids under pressure. The torque converter is an example of a device which operates according to the principles of hydrodynamics. The automatic transmission valve body, clutches, and servos depend on the laws of hydrostatics. The automatic transmission technician need not be a hydraulic engineer, but he must understand the several principles which apply directly to the operation of an automatic transmission. These principles are presented in this unit.

Liquids Cannot Be Compressed

Hydraulic devices are able to function because liquids resist compression. If a cork were forced into a bottle completely

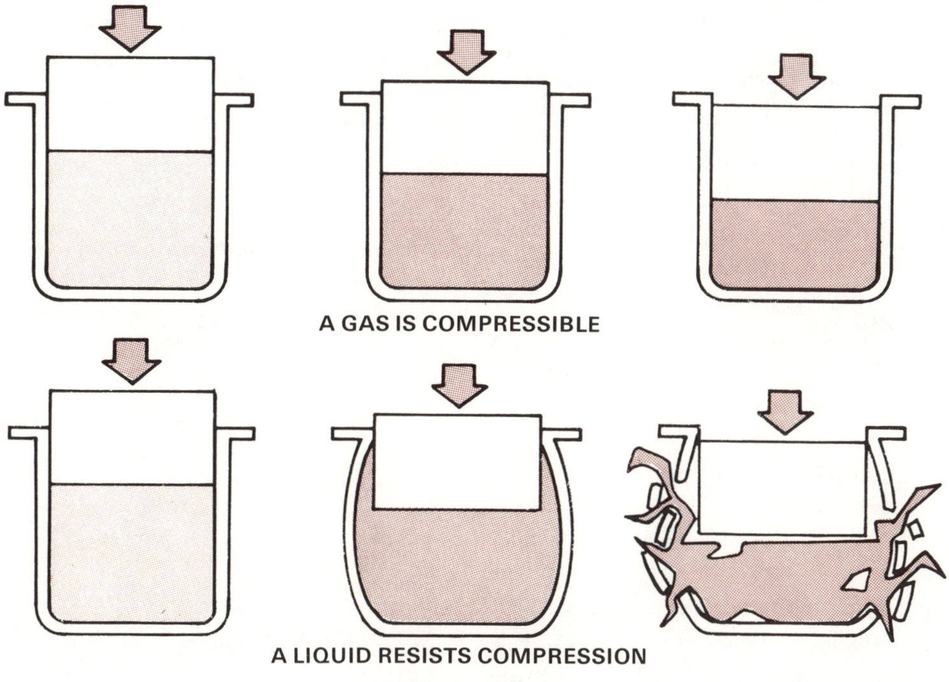

Figure 7-1

Liquids Cannot Be Compressed
(Vickers Div., Sperry Rand Corp.)

full of water, eventually the cork would stop moving. If the cork were forced to move further, the bottle would break. This characteristic of fluids to resist compression allows them to transmit motion and pressure.

Liquids Can Transmit Motion

Since liquids are not compressible they may be used to transmit motion. If a cylinder contains two pistons separated by a liquid, pushing on piston "A" will

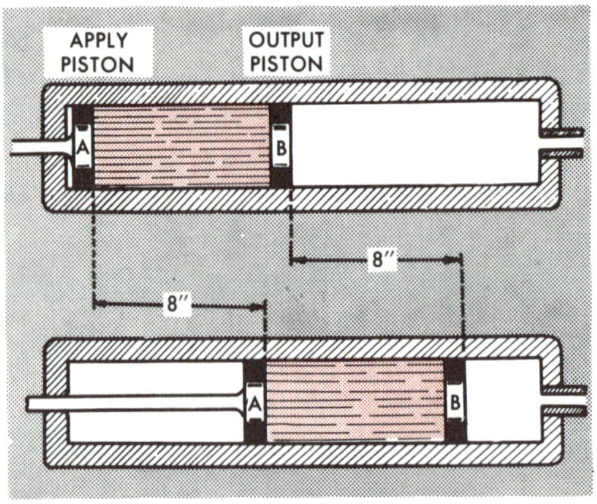

Figure 7-2

Liquids Can Transmit Motion
(General Motors Corp.)

cause movement at piston "B." Because piston "A" starts the movement, it is called the apply piston. Piston "B" is called the output piston. If the apply piston is moved eight inches the output piston will also move eight inches. This demonstrates that motion may be transmitted by a liquid.

The same principle may be used to transmit motion from one cylinder to another. In figure 7-3, two cylinders are connected by a pipe. When the apply piston in cylinder "A" is moved four inches, the output piston in cylinder "B" also moves four inches. This principle is, of course, also used in hydraulic brake systems where a master cylinder initiates a push which is transmitted to each of the vehicle's wheel cylinders.

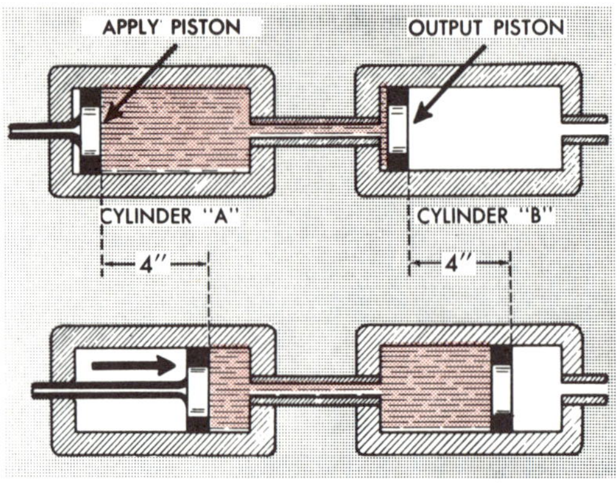

Figure 7-3

Liquids Transmit Motion between Cylinders
(General Motors Corp.)

A Liquid Under Pressure Transmits Pressure Equally in All Directions

Pressure is exerted on a confined liquid by applying a force to some area in contact with the fluid. For example, in figure 7-4 a piston is used to create a pressure in the fluid. If 100 pounds of force is applied to a piston with a one square-inch area, a pressure of 100 pounds per square inch (abbreviated psi) is generated. The important thing to understand here is that the 100 psi is available everywhere in the fluid. If a

Figure 7-4

Pressure Is Transmitted in All Directions Equally
(General Motors Corp.)

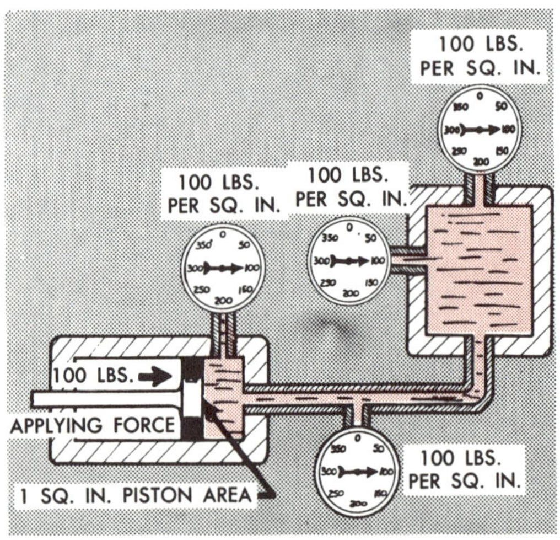

pressure gage were installed anywhere in the system it would read 100 psi. This principle was formulated by a Frenchman, Blaise Pascal, in the seventeenth century, and is referred to as Pascal's Law. Pascal's Law states: Pressure at any one point in a static liquid is the same in every direction and exerts equal force on equal areas.

Pressure Equals Force Divided by Area

Pressure in a closed system may be calculated by dividing the force applied by the area of the apply piston. In figure 7-5 a 100-pound force is applied to a piston with a one square-inch area. Using the principle Pressure equals Force divided by Area (P = F/A), 100 is divided by 1, which equals 100. Since 100 is the pressure per square-inch, it is expressed as 100 psi. If the same force is applied to a piston with a two square-inch area, the resulting pressure would be 50 psi.

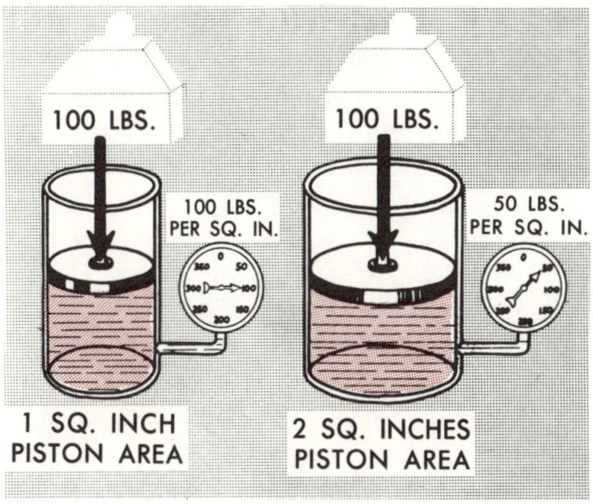

Figure 7-5

Pressure Equals Force Divided by Area
(General Motors Corp.)

Leverage Can Be Gained Hydraulically

One of the most significant results of Pascal's work was the discovery that fluids might be used to gain leverage. If two cylinders of different diameters are connected as shown in figure 7-6 with a fluid trapped between them, a curious thing happens. A two-pound force on the piston with the one square-inch area develops a pressure in the system of two psi (P = F/A; 2 = 2/1). This two psi is available everywhere in the system. When this two psi is distributed across a piston with a 50 square-inch area, the output force is 100 pounds. In other words, the two-pound weight on the small piston balances a 100-pound weight on the larger piston.

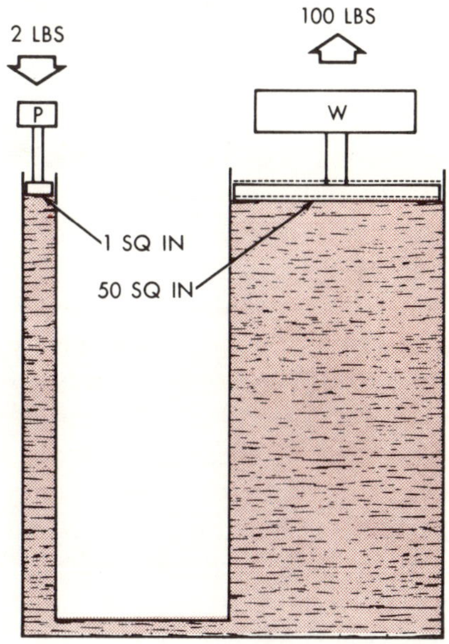

Figure 7-6

Leverage Can Be Gained Hydraulically
(Vickers Div., Sperry Rand Corp.)

All mechanics have used a hydraulic jack to lift a vehicle. This is the principle operating in the jack. A man is able to lift a 4,000 pound vehicle with his own weight through hydraulic leverage. Leverage is gained, however, at a loss of distance. In order to move the larger piston in figure 7-6 a distance of one inch, the smaller piston must be moved a distance of 50 inches. Using the example of the hydraulic jack again, the jack handle must move a greater distance than the vehicle being lifted.

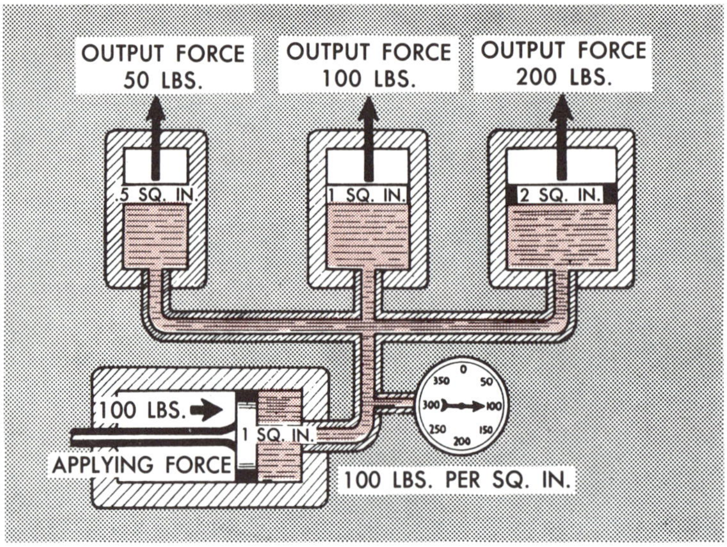

Figure 7-7

Output Force Equals Pressure Times Area
(General Motors Corp.)

Output Force Equals Pressure Times Area

The principles of hydraulic leverage may be applied to a system with any number of cylinders. In figure 7-7, a system of four cylinders is shown. The apply piston has a one square-inch area. A force of 100 pounds on the apply piston develops a pressure in the system of 100 psi. Since each of the output pistons is a different size, there is a different force at each cylinder. The output may be calculated by multiplying the pressure in the system times the area of the output piston.

The system pressure in figure 7-7 is 100 psi. The output force at the .5 square-inch area piston is 50 pounds ($100 \times .5 = 50$); at the one square-inch area piston it is 100 pounds ($100 \times 1 = 100$); and at the two square-inch area piston, the force is 200 pounds ($100 \times 2 = 200$). The larger the output piston the larger the output force. A common application of this principle is the hydraulic brake, which uses a small master cylinder piston to apply pressure to a larger piston at the wheel cylinder, gaining a good deal of hydraulic leverage.

Flow Is Created by a Pressure Difference

Flow is the movement of fluid in a hydraulic system. In order to have flow there must be a difference in pressure between two points. When the water tap is turned on at the kitchen sink, water flows out. This is because the water system has built up a pressure in the water pipes. When the tap is opened, water moves from the high-pressure area to the low-pressure area. When liquid is not subject to a pressure, it seeks a level, as shown in figure 7-8.

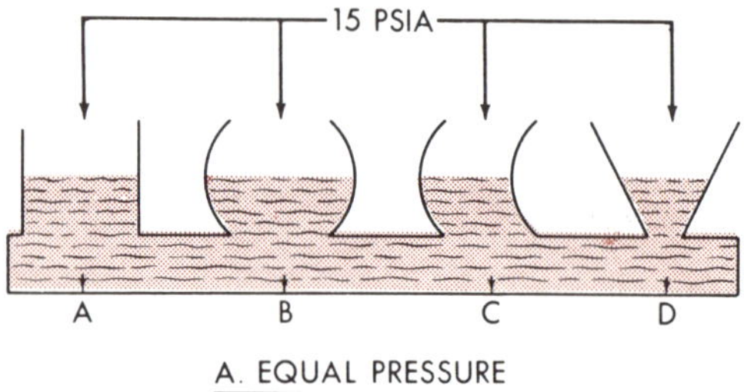

Figure 7-8

Fluid Not Subject to Pressure Seeks a Level
(Vickers Div., Sperry Rand Corp.)

Notice that the vessel in figure 7-8 has four differently shaped chambers. The size and shape of the chambers, however, does not affect the fluid level. If a piston were forced into one of the chambers, a pressure would be created. Fluid would move—or flow—from the chamber with

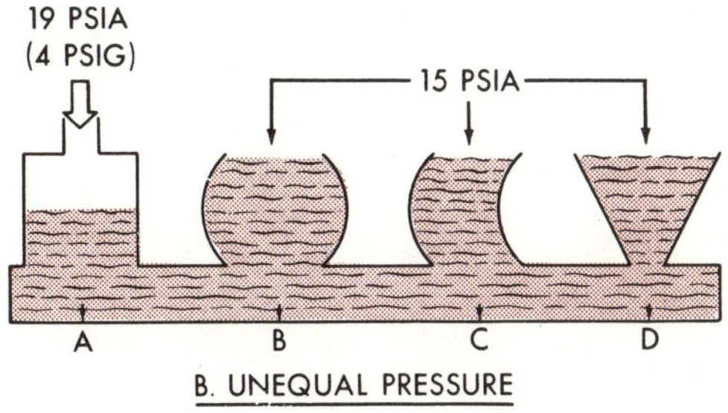

Figure 7-9

Pressure Difference Causes Fluid to Flow
(Vickers Div., Sperry Rand Corp.)

the pressure build-up to the chambers with lower pressures. In figure 7-9 pressure in one chamber has caused fluid to flow into the other three. The fluid level has risen in the three chambers. If the piston were to continue moving, fluid would continue to flow.

Flow through an Orifice

Fluid in an automatic transmission is routed through many holes and passageways on its way to operate the various hydraulic devices. As fluid moves through a hole or orifice there is a difference in pressure on each side of the orifice. The fluid is restricted on one side of the orifice and the pressure rises there.

If the flow is blocked, then according to Pascal's Law, pressure will be the same in all parts of the system. Thus, both sides of the orifice register equal pressure as shown in figure 7-11.

These are the hydraulic principles which go to work in an automatic transmission. The application of these principles will be the subject of study in units which follow.

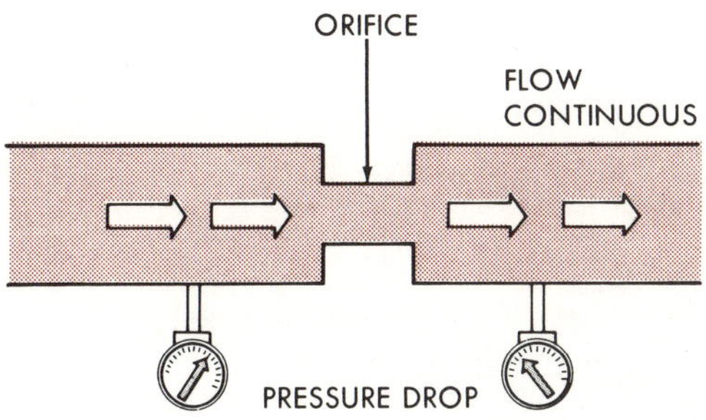

Figure 7-10

Continuous Flow across an Orifice Causes a Pressure Drop
(Vickers Div., Sperry Rand Corp.)

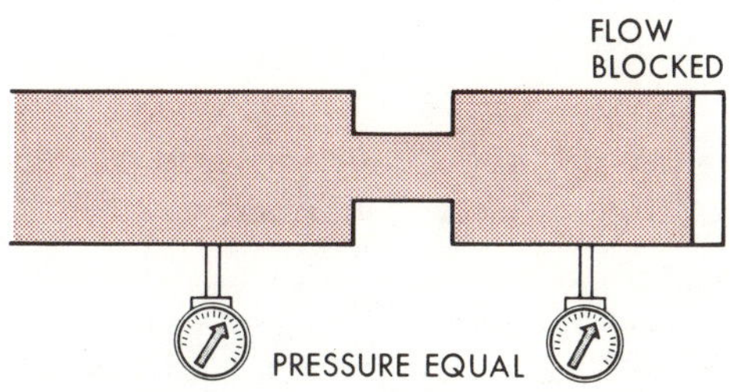

Figure 7-11

Equal Pressure when Flow Is Blocked
(Vickers Div., Sperry Rand Corp.)

UNIT 7
Principles of Hydraulics

True-False Questions (Indicate the correct choice)

1. Hydrostatics is the study of moving liquids. T F

2. Liquids cannot be compressed. T F

3. Liquids can transmit motion. T F

4. A liquid under pressure exerts force equally in all directions. T F

5. Fluids cannot be used to gain leverage. T F

6. Flow is created by a pressure difference. T F

7. When oil flows through an orifice there is no difference in pressure on either side of the orifice. T F

Problems

1. Piston A and piston B below are separated by fluid. If piston A is moved four inches, what will happen to piston B?

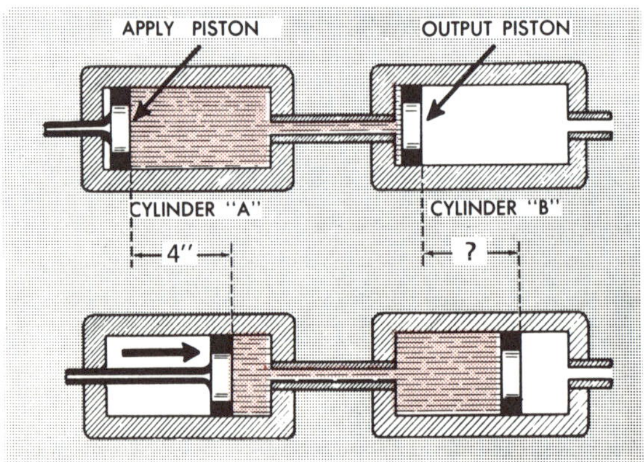

2. Using the formula *Output Force Equals Pressure Times Area,* calculate the output force in each of three output cylinders.

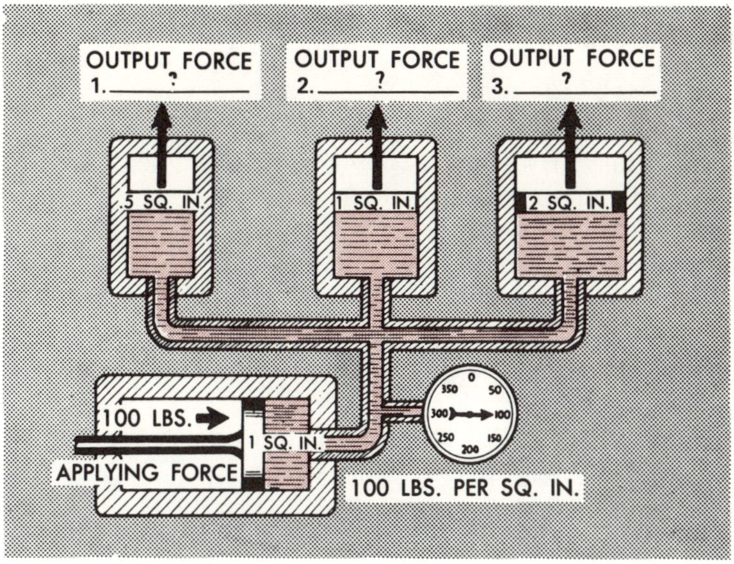

1.____________________ 2.____________________ 3.____________________

UNIT 7
Principles of Hydraulics

True-False Questions

1. This statement is false. Hydrostatics is the study of liquids under pressure. The study of moving liquids is called hydrodynamics.

2. This statement is true. Liquids resist compression.

3. This statement is true. Since liquids are not compressible they may be used to transmit motion from one place to another.

4. This statement is true. Pascal's Law states that pressure at any one point in a static liquid is the same in every direction and exerts equal force on equal areas.

5. This statement is false. One of the most significant results of Pascal's work was the discovery that fluids *could* be used to gain leverage. The hydraulic jack is an example of this principle.

6. This statement is true. In order to have flow there must be a difference in pressure between two points.

7. This statement is false. When oil flows through an orifice there will be a pressure difference between the two sides of the orifice.

Problems

1. Piston B would move four inches as shown in the sketch below. This demonstrates how liquids can be used to transmit motion.

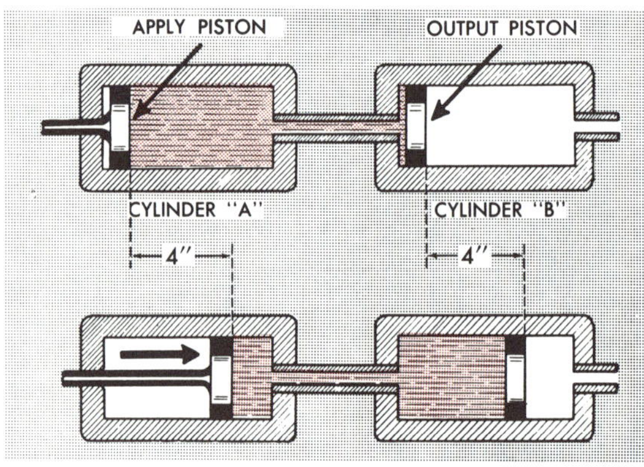

2. The output force for each of the three output pistons is:

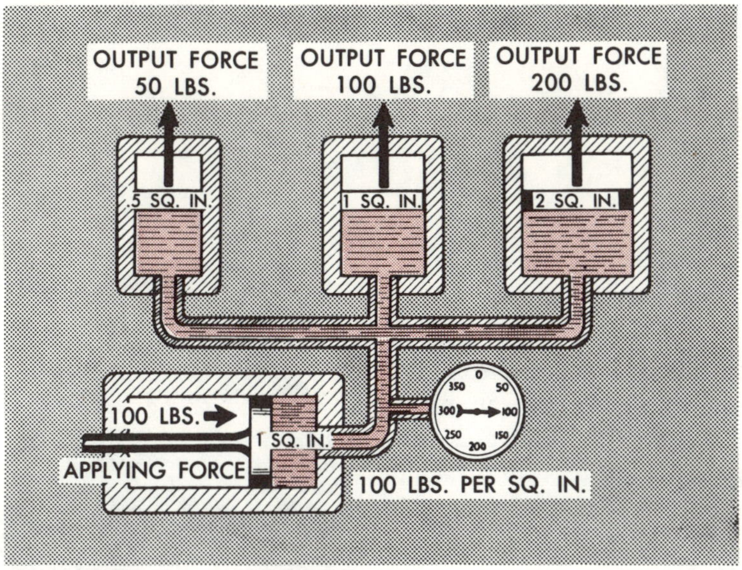

1. 50 lbs.

2. 100 lbs.

3. 200 lbs.

Pumps for Automatic Transmissions

A pump is required in an automatic transmission to push on the hydraulic fluid and create flow and pressure. The pump's job is to convert mechanical energy from the engine into pressure energy to be used in the transmission. The pressure generated by the pump will be used to sense vehicle gearing requirements and effect a gear change in the planetary gearbox. In addition, the pump circulates fluid throughout the transmission for cooling and lubrication.

Number and Location of Pumps

Automatic transmissions may be equipped with one or two pumps. The location of the pumps is shown in figure 8-1. All automatic transmissions use a front pump. The front pump is driven by the engine, usually through two dogs on the end of the torque converter housing. The front pump operates whenever the engine is running.

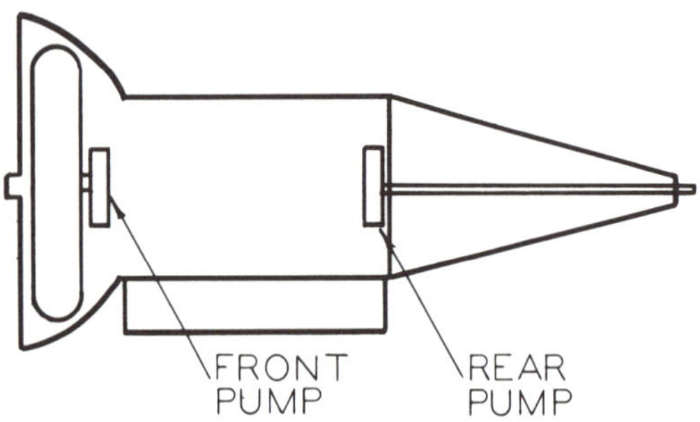

Figure 8-1

Number and Location of Pumps

Figure 8-2

Pump Drive on Converter Housing

The rear pump, if the transmission uses one, is connected to and driven by the output shaft. The rear pump operates whenever the output shaft revolves, so the rear pump can pump only when the vehicle is moving. In transmissions equipped with two pumps, the front pump may do the whole job of supplying oil, with the rear pump held in reserve. More commonly, the front pump is used until the vehicle reaches a predetermined road speed, when the rear pump is then utilized.

In the late 1950s and early 1960s,

manufacturers began to remove the rear pump from their transmissions. Using a single pump offered several advantages. A single pump system costs less and, of course, has fewer components to wear out or fail. Since pumps require power to drive them, a single pump system uses less of the engine's power.

With these advantages, however, came some disadvantages. A car without a rear pump in the transmission cannot be pushed to be started. Pushing a vehicle with a rear pump rotates the drive shaft, which in turn rotates the transmission's output shaft. Rotating the output shaft causes the rear pump to deliver oil. Oil pressure can then be used to put the transmission in gear. Without a rear pump, no oil pressure can be generated. Thus, the transmission cannot be put in gear and the vehicle cannot be push-started. Another problem with a single pump system is related to towing the vehicle. Towing a vehicle without a rear pump means that (if the rear wheels are on the ground) the transmission components are being rotated by the drive shaft but there is no oil circulating for lubrication. Most operator's manuals call for the vehicle to be towed with the rear wheels off the ground or the drive shaft disconnected to prevent wear of parts due to a lack of lubrication.

Types of Pumps

Several types of pumps have been used in automatic transmissions. The earliest automatic transmissions used an *external gear pump.* An external gear pump consists of two gears with external teeth meshed together in a housing (figure 8-3). One of the gears in the pump is driven which, in turn, drives the other. As the gear teeth separate and travel past an inlet port, a partial vacuum is created. Fluid is then drawn into the inlet port and carried around in the teeth of the gears. As the gear teeth come together at the outlet, there is no place for the fluid to go but out.

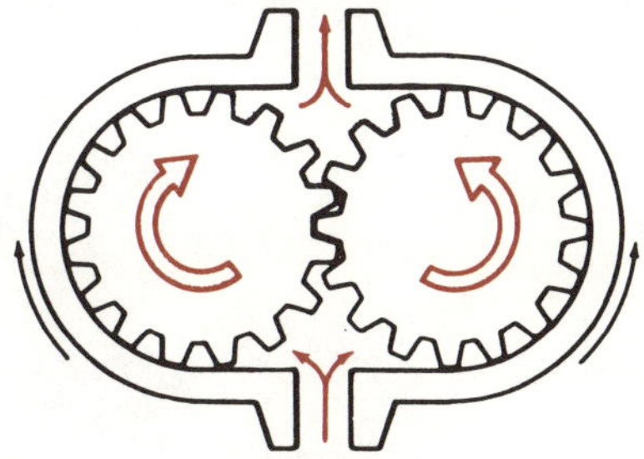

Figure 8-3

External Gear Pump
(Vickers Div., Sperry Rand Corp.)

External gear pumps are used very often as oil pumps for engines, but they have become less popular for automatic transmissions. An external gear pump has a fluctuating delivery. This is objectionable because a smooth flow of oil is necessary in an automatic transmission.

Another pump used in automatic transmissions is the *vane pump.* The vane pump uses a series of vanes which sweep an eccentric cavity. When the pump is driven, the unit holding the vanes rotates. The vanes fit tightly against the walls of the pumping chamber. Fluid enters a low-pressure area created where the eccentric cavity begins, and is carried around by the vanes to the outlet port. The vane pump provides a smooth flow of fluid but because it is composed of a large number of parts, it is expensive to produce. (See figure 8-4, next page.)

The most popular pump for automatic transmissions is the *crescent pump.* This pump is also called an *internal pump,* or an *IX* (for internal external) *pump.* The crescent pump consists of an inner gear keyed to the drive mechanism, a larger external gear, and a crescent seal installed in a close fitting housing.

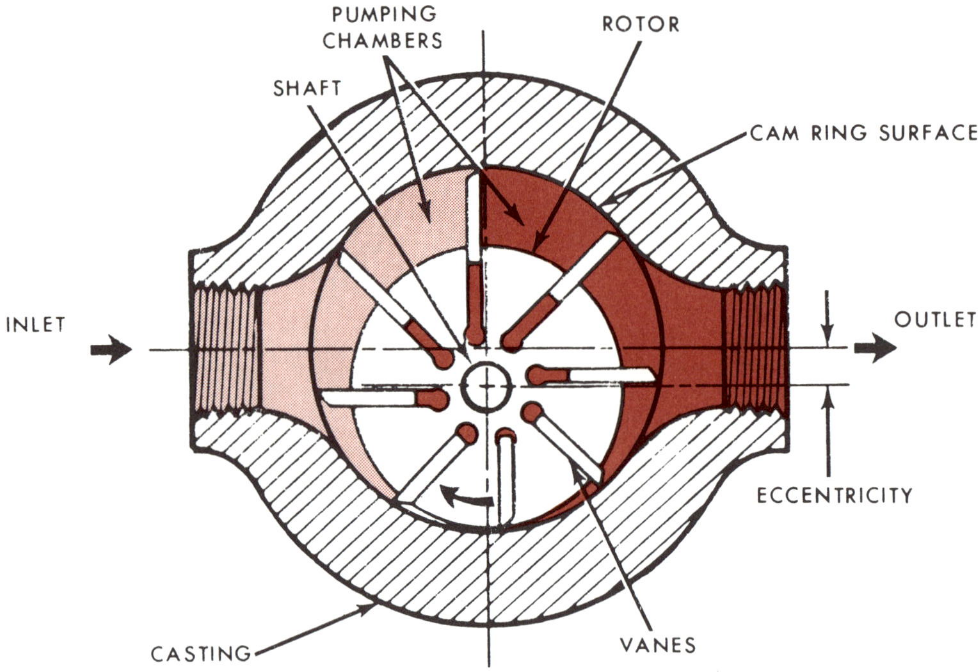

Figure 8-4

Vane Pump
(Vickers Div., Sperry Rand Corp.)

The two gears are not concentric, so as they rotate past the inlet and unmesh, a low-pressure area is created. Fluid is pulled in and carried around in the teeth to the outlet. The crescent seals the inlet port from the outlet port. At the outlet port the fluid has no place to go but out. Most automatic transmissions now use a crescent pump. The fluid delivery from a crescent pump is smooth, the pump is compact, and it may easily be designed to fit around an input or output shaft of an automatic transmission.

Figure 8-5

Crescent Pump

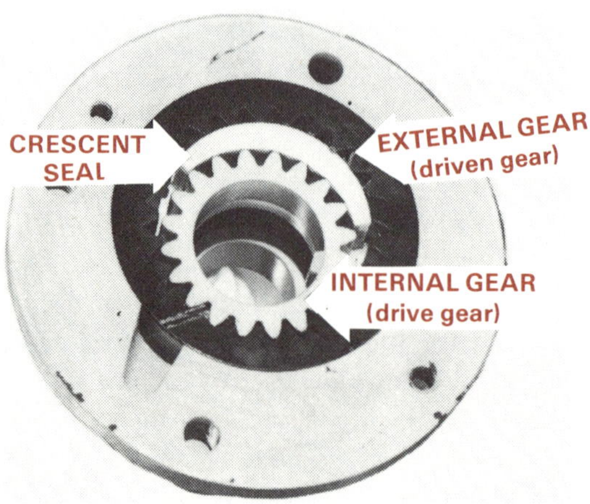

Pump Terminology

Several terms are used to describe pumps. The more common pump terms and what they mean are described below.

Non-positive Displacement. If a pump's inlet and outlet are connected hydraulically so that fluid can be recirculated in the pump, it is called a non-positive displacement pump. Automatic transmissions do not use this kind of pump. An example of a non-positive displacement pump is the water pump on the automobile engine.

Positive Displacement. If the inlet of a pump is sealed from its outlet, the pump delivers to the outlet all the fluid that is available at the inlet. This kind of pump is classified as a positive displacement pump. It must have a relief or pressure regulation valve to protect it from excessive pressure. All automatic transmission pumps are classified as positive displacement pumps.

Flow Rating. Pumps are rated according to how many gallons per minute (gpm) they can deliver. Pumps for automatic transmissions must have a high flow rating in order to supply the demand for fluid to operate the controls, as well as to cool and lubricate.

Pressure Rating. Pumps produce flow. The resistance to this flow results in a pressure. A pump's pressure rating is the amount of pressure a pump can withstand for a given time without damage. An automatic transmission pump must be able to withstand pressures in excess of 300 pounds per square inch.

Capacity Rating. The amount of flow available at a given pressure and rpm.

UNIT 8
Pumps for Automatic Transmissions

True-False Questions (Indicate the correct choice)

1. The pump is required to push on the hydraulic fluid and create the fluid flow to be used in the transmission. T F

2. All automatic transmissions are equipped with two oil pumps. T F

3. The front pump of an automatic transmission usually is driven by the torque converter and pumps whenever the engine is running. T F

4. The rear pump of an automatic transmission is driven by the output shaft and pumps whenever the vehicle is moving. T F

5. A vehicle using an automatic transmission without a rear pump can be push-started. T F

6. A vehicle using an automatic transmission without a rear pump can be towed at high speed with its rear wheels on the ground. T F

7. One of the problems with an external gear pump is a fluctuating delivery. T F

8. The most popular type of pump for automatic transmissions is the vane pump. T F

9. If a pump's inlet and outlet are connected hydraulically so fluid can be circulated in the pump, it is a positive displacement pump. T F

10. Pumps produce flow, not pressure. T F

Multiple Choice Questions (Indicate the correct choice)

1. Oil flow created by the pump in an automatic transmission is used to:
   a. sense vehicle gearing requirements
   b. effect gear changes
   c. cool and lubricate
   d. all of the above

2. One of the advantages of using a single pump in the automatic transmission is:
   a. the vehicle can be push-started
   b. the vehicle can be towed
   c. less horsepower is consumed in pump operation
   d. less oil pressure is required

3. The most common pump found in automatic transmissions is:
   a. crescent   b. vane   c. external gear   d. piston

4. The advantage of a crescent pump over the other types of pumps is:
   a. the unit is compact
   b. the unit may be designed to fit around an input or output shaft (concentric)
   c. the oil delivery is smooth
   d. all of the above

5. The amount of oil a pump can move in a given time is expressed as its:
   a. displacement ratio   b. flow rating   c. pressure rating

Parts Identification

Identify the numbered components of the crescent pump shown.

1.____________________ 2.____________________ 3. ____________________

UNIT 8
Pumps for Automatic Transmissions

True-False Questions

1. This statement is true. The pump is turned by the engine and creates fluid flow for the transmission.

2. This statement is false. An automatic transmission may be equipped with either one or two pumps.

3. This statement is true. The front pump is driven by the torque converter and pumps whenever the engine is running.

4. This statement is true. The output shaft of an automatic transmission drives the rear pump whenever the vehicle is moving.

5. This statement is false. If an automatic transmission does not have a rear pump it cannot be push-started because without a rear pump, oil cannot be circulated through the transmission unless the engine is running.

6. This statement is false. A vehicle using an automatic transmission with no rear pump should not be towed at high speed with its rear wheels on the ground since towing would cause the transmission components to rotate but would not provide oil to lubricate them.

7. This statement is true. External gear pumps are unable to deliver the smooth flow of oil required in automatic transmissions.

8. This statement is false. Most automatic transmissions use a crescent pump, rather than the vane pump.

9. This statement is false. If a pump's inlet and outlet are connected together hydraulically so fluid can be recirculated in the pump, it is a non-positive displacement pump.

10. This statement is true. Pumps provide fluid flow; pressure is due to the resistance to this flow.

Multiple Choice Questions

1. The correct choice is *d.* The oil flow created by the pump in an automatic transmission is used to sense the vehicle's gearing requirements and effect gear changes, as well as to cool and lubricate.

2. The correct choice is *c.* Manufacturers went to a single pump because it took less power from the engine.

3. The correct choice is *a.* Most automatic transmissions use a crescent-type pump.

4. The correct choice is *d.* The crescent pump is compact, concentric in design, and provides a smooth oil delivery.

5. The correct choice is *b.* Flow rating is the amount of oil a pump can deliver in a given time.

Parts Identification

1. The main components of a crescent oil pump are:

1. Crescent Seal   2. External Gear   3. Internal Gear

Valves

The control system in an automatic transmission is a network of valves. All of these valves respond to the laws of hydraulics. The purpose of this unit will be to explain the principles which are common to all the various valves in the valve body. Hydraulic valves can be divided into two general groups: those that *regulate* flow and pressure, and those that *direct* flow and pressure.

Valves that Regulate Flow and Pressure

One means of regulating flow and pressure is the *orifice*. An orifice is simply a restriction in a hydraulic passageway. It acts to slow down fluid flow by creating a back pressure or delaying a pressure build-up. When fluid is pumped through an orifice, there is not room for all of it to go through at once. A back pressure is created on the pump side. As long as there is a flow through the orifice, a pressure difference is maintained at the orifice (figure 9-1). An automatic transmission control system or valve body contains many restricted passages which act as orifices.

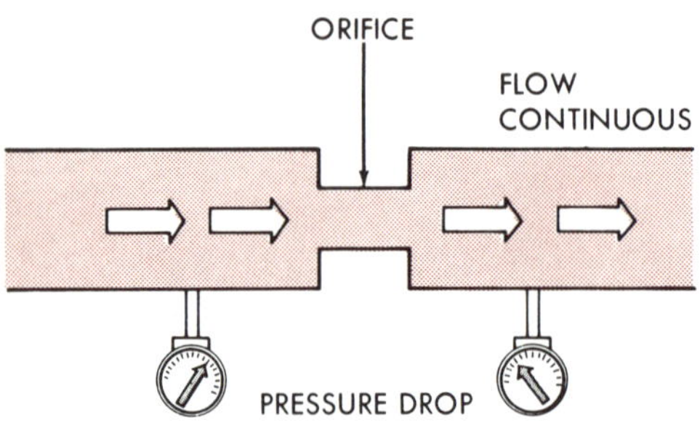

Figure 9-1

An Orifice Regulates Flow
(Vickers Div., Sperry Rand Corp.)

Another valve that is used to regulate flow in an automatic transmission is the *non-return* or *check valve*. How this valve works is shown in figure 9-2. The non-return or check valve consists of a small steel ball located in a fluid passageway and allows fluid to move in one direction only. In the open position, fluid pushes the ball against a stop and moves around it. If fluid should try to go in the opposite direction, the ball is pushed into a position to block fluid flow. Many automatic transmissions use several non-return valves in the control system or valve body.

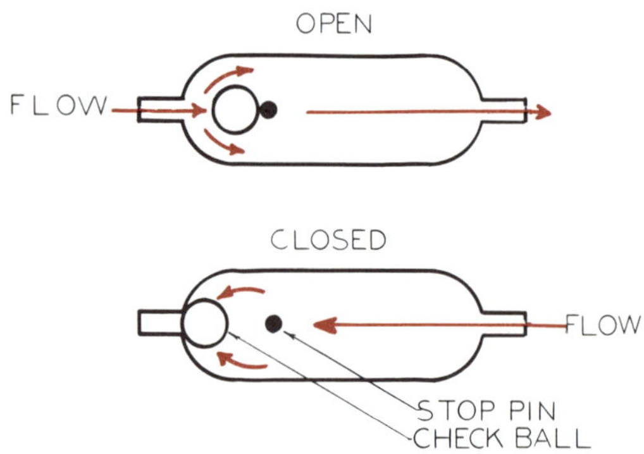

Figure 9-2

Operation of a Non-Return Valve

The *pressure regulator valve* is also important because it regulates the pressure from the pump. The pump in the automatic transmission delivers a fluid at a rate determined by how fast it is driven. This means that the faster the pump turns, the more flow it provides and the more pressure it generates. The pressure regulator operates to hold the pressure in the system at a desired level, regardless of how fast the pump is driven beyond the minimum speed needed.

A typical pressure regulator is shown in figure 9-3. The pressure regulator is located close to the pump, and the fluid flow from the pump is routed to the regulator valve. The regulator consists of a cylindrically shaped valve which is positioned in a chamber by a calibrated spring. Fluid pressure from the pump acts on the valve and spring. When the pressure rises above the required amount, the valve is forced over, uncovering a passageway back to the oil sump. Some fluid from the pump is allowed to go back to the sump. Every time pump pressure rises, the valve moves to bleed off the excess pressure. This action maintains a constant pressure at the outlet. The amount of pressure at the outlet is determined by the tension of the calibrated spring. If the spring tension were increased the pressure would be increased because it would take more pump pressure to move the valve into the bleed-off position. If the spring tension were decreased the pressure would decrease.

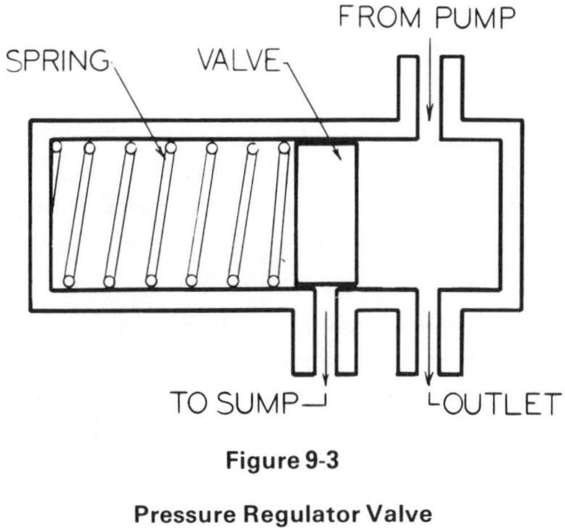

Figure 9-3

Pressure Regulator Valve

Many automatic transmissions use a type of pressure regulator valve called a *variable pressure regulator.* A variable pressure regulator provides different pressures in the transmission for different gear shifts. For example, to hold the transmission in low range it may take only 90 psi. Drive range, however, may require 150 psi, and reverse as much as 350 psi. A variable pressure regulator is shown in figure 9-4. The variable pressure regulator acts just as the regulator

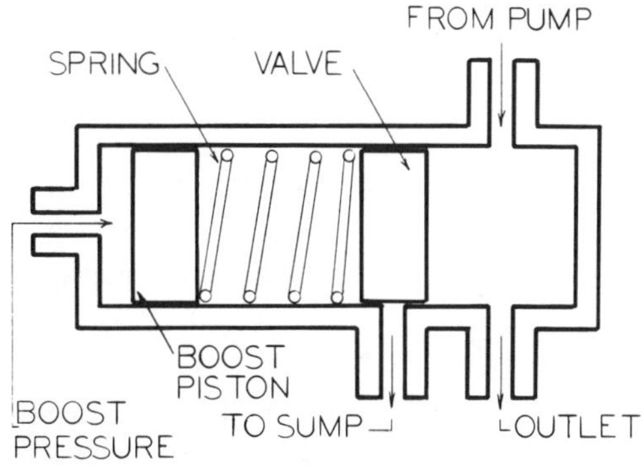

Figure 9-4

Variable Pressure Regulator Valve

shown in figure 9-3, except that some provision is made for changing the tension on the calibrated spring. The spring tension is changed by a piston which pushes against it. When pressure beyond the spring setting is required, pressure is directed to the piston. The piston is forced against the spring, increasing its tension. This action prevents the regulator valve from bleeding off pressure, and results in a high pump pressure. The pressure that acts on the boost piston to raise the spring force and pump pressure is generally known as boost pressure. When less pressure is required, less fluid is directed at the boost piston. The spring is allowed to relax and the regulator valve can move to bleed off pressure at a lower level.

Valves That Direct Flow

Spool valves are used to direct flow in an automatic transmission. As figure 9-5 shows, a spool valve is cylindrically shaped, with two or more lands, and with grooves or valleys between the lands.

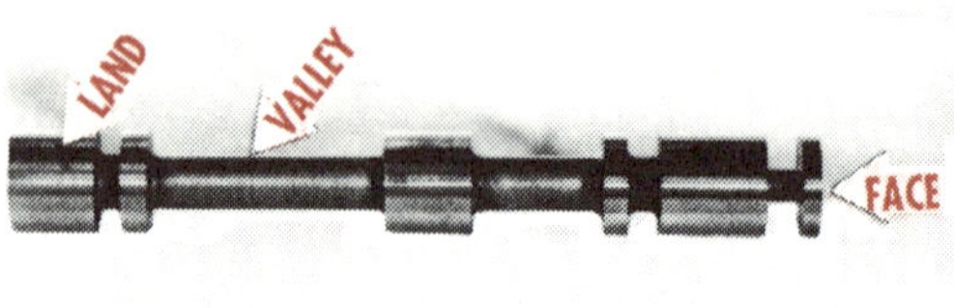

Figure 9-5

Spool Valve

The spool valve is closely fitted to a round bore, and slides in the bore on a pressurized film of liquid. Fluid passages are opened or closed depending on the valve land positions. The fluid directed at the valve is allowed to flow past the valve and enter another passageway only if the valve is positioned so as not to obstruct the flow. In figure 9-6 fluid is directed to

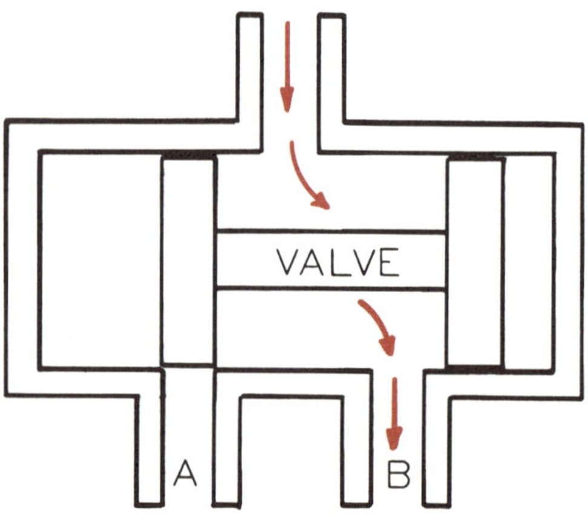

Figure 9-6

Spool Valve Directing Flow

the spool valve which is positioned to allow fluid to flow into passageway "B." If the valve were moved to the left, passageway "B" would be blocked and passageway "A" would be opened. Thus, the spool valve is able to direct flow. An automatic transmission spool valve typically has many lands, and is responsible for directing flow into many different passageways.

The spool valve must be moved back and forth in its bore to direct flow into different passageways or circuits. Three different methods are used to position the spool valves—manual linkage, hydraulic pressure, and spring pressure. Figure 9-7 shows how a spool valve may be positioned with manual linkage. If a lever is connected to the spool valve as shown,

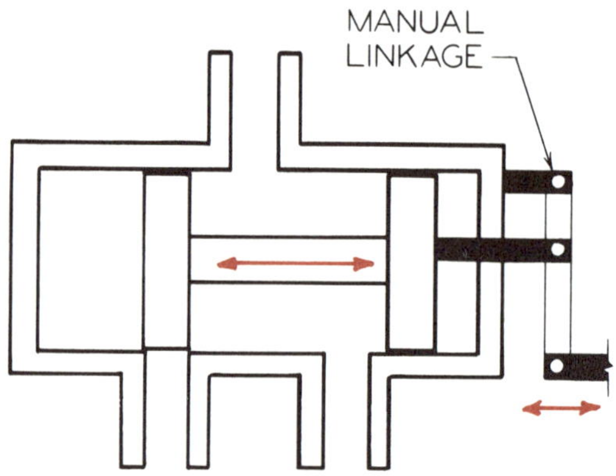

Figure 9-7

Spool Valve Positioned by Manual Linkage

movement of the lever would cause movement of the spool valve. This, in turn, would change the direction of flow from one circuit to another. In an automatic transmission, several spool valves are connected to a manual linkage so that the driver may select the circuit he wishes.

Spring pressure and hydraulic pressure are often used together to position a spool valve. Figure 9-8 shows a spool valve being positioned by spring pressure and hydraulic pressure. The spring operates to hold the spool valve in a position to close off a passageway. Hydraulic pressure directed at the other end of the valve provides an opposing force to the spring. If the hydraulic pressure on the end of the spool valve increases, at some

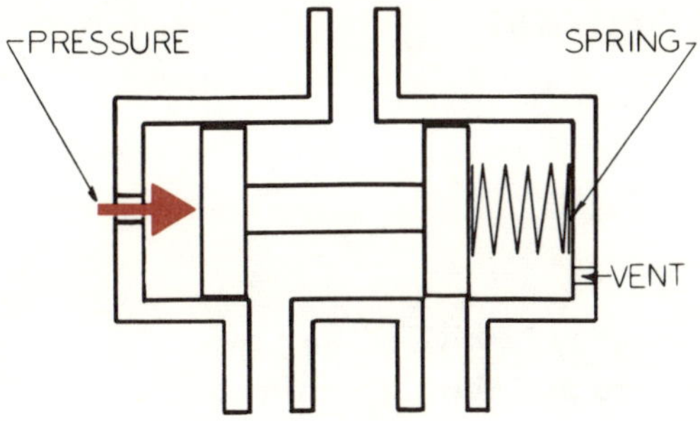

Figure 9-8

Hydraulic and Spring Pressure Used to Position a Spool Valve

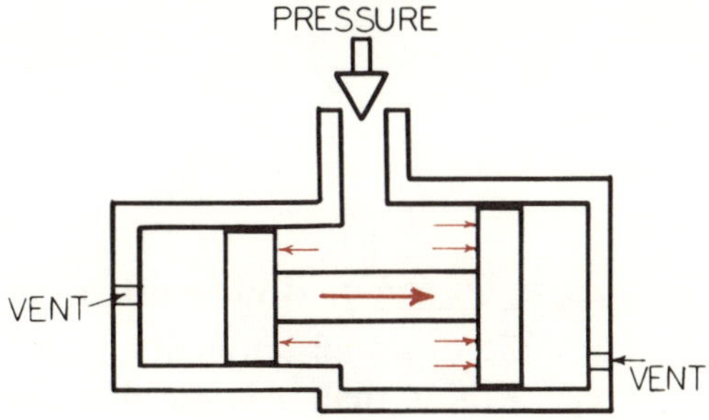

Figure 9-9

Spool Valve Positioned by Different Sized Lands

point it will exceed the spring pressure. When this happens, the spool valve will be moved to the right, compressing the spring. The fluid directed at the valve is allowed to flow past the valve and enter the other passageway. If the hydraulic pressure is decreased, the spring will act to reposition the valve in its original location.

Hydraulic pressure alone may be used to move a spool valve. When two adjacent lands of a spool valve have different diameters and a pressure is applied between them, the larger land receives a harder push. This compares to different sized output pistons in a hydraulic system as discussed in Unit 7. Figure 9-9 shows a valve with two different sized lands. When pressure is applied, a larger force acts on the larger land. Thus, the valve moves to the right. Most of the spool valves in an automatic transmission have different sized lands so that hydraulic pressure may be used to position them.

UNIT 9
Valves

True-False Questions (Indicate the correct choice)

1. Hydraulic valves for automatic transmissions may be divided into two groups—those that *regulate* flow and pressure and those that *direct* flow and pressure. T F

2. An orifice is an example of a valve that *directs* flow and pressure. T F

3. When fluid is pumped through an orifice, a back pressure is created on the pump side. T F

4. A steel ball located in a fluid passageway is called a non-return valve. T F

5. The pressure regulator valve operates to hold the pressure in the system at a constant amount. T F

6. If the pressure regulator spring tension were increased the system pressure would decrease. T F

7. Most automatic transmissions use a type of pressure regulator valve called a variable pressure regulator. T F

8. Spool valves are used to *direct* flow in an automatic transmission. T F

9. Spool valves may be positioned by manual linkage. T F

10. When two adjacent lands of a spool valve have different diameters and a pressure is applied between them, the larger land receives a harder push from the pressure. T F

Parts Identification

Identify the components of a spool valve, shown below.

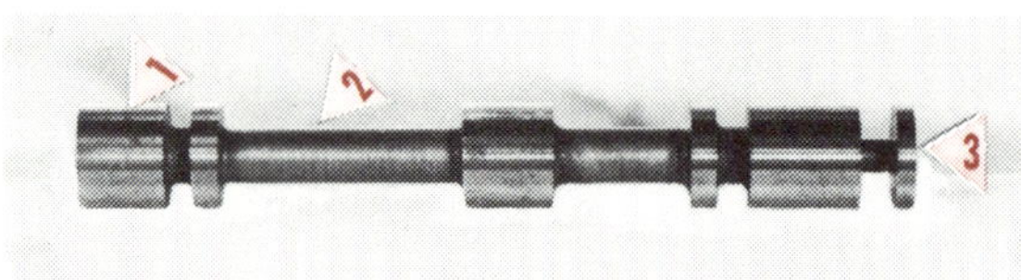

1.________________ 2.________________ 3.________________

<u>*Multiple Choice Valve Problems*</u> (Indicate the correct choice)

What will happen to the spool valves pictured below when pressure is applied?

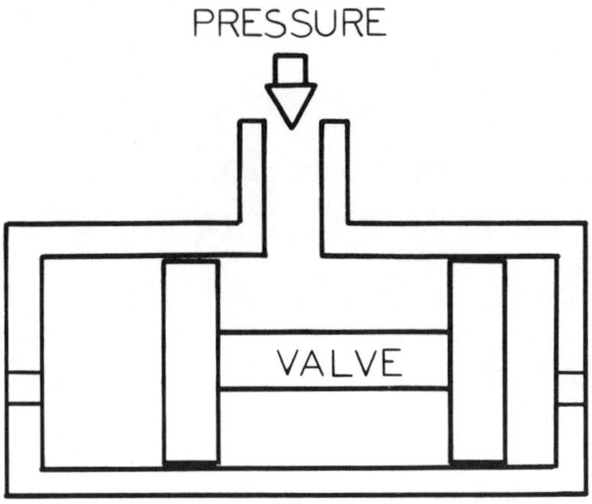

1. The valve will:
   a. move to the left
   b. move to the right
   c. not move

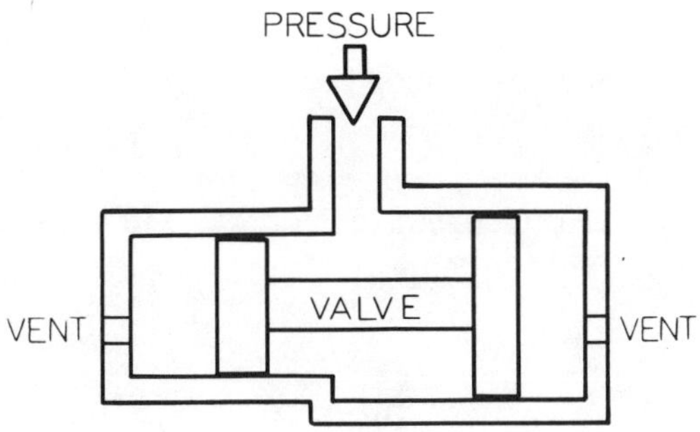

2. The valve will:
   a. move to the left
   b. move to the right
   c. not move

UNIT 9
Valves

True-False Questions

1. This statement is true. The two general groups into which valves may be divided are those that regulate or control flow and pressure, and those that direct flow and pressure.

2. This statement is false. An orifice is used to *regulate* flow and pressure, not *direct* flow and pressure.

3. This statement is true. When fluid is pumped through an orifice, there is not room for all of it to go through at once. Thus, back pressure is created on the pump side.

4. This statement is true. A non-return valve consisting of a steel ball is used to regulate flow in one direction.

5. This statement is true. The pump delivers fluid at a rate determined by how fast it is driven. The pressure regulator holds the pressure at a constant amount regardless of how fast the pump is driven.

6. This statement is false. If the spring tension were increased, the pressure would be increased because it would take more pump pressure to move the valve into the bleed-off position.

7. This statement is true. Since different pressures are required for different shifts, a variable pressure regulator is required.

8. This statement is true. Spool valves are used to *direct* flow in an automatic transmission.

9. This statement is true. Several methods are used to position spool valves, including manual linkage, hydraulic pressure, and spring pressure.

10. This statement is true. When pressure is applied, a larger force acts on the larger land, causing movement. This is how hydraulic pressure is used to position a spool valve.

Parts Identification

The parts of a spool valve are:

1. Land
2. Valley
3. Face

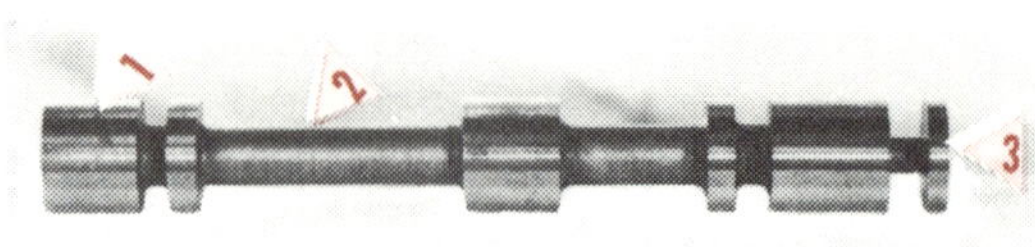

Multiple Choice Valve Problems

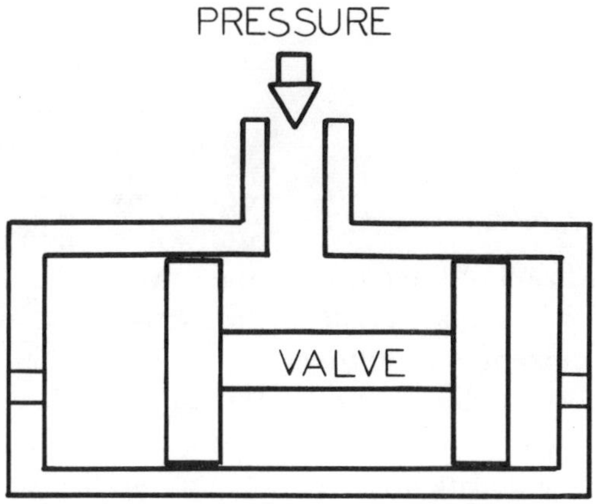

1. The correct choice is *c.* Since the valve has equal sized lands, the forces are equalized and the valve will remain stationary.

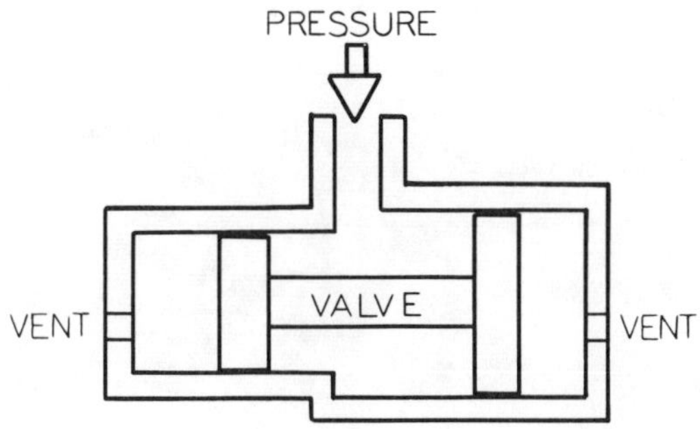

2. The correct choice is *b.* The larger land on the right has a larger force acting on it, so the valve will move to the right.

The Valve Body

The valve body is the brain of the automatic transmission. Its job is to sense the vehicle's gearing requirement and automatically select the correct gear ratio. The valve body is made up of a number of spool valves contained in a housing. Each of the valves has a specific job to do. This unit will explain the operation of each of the valves typically found in an automatic transmission.

Pressure Regulator Valve

The fluid flow into the valve body begins at the sump or oil pan on the

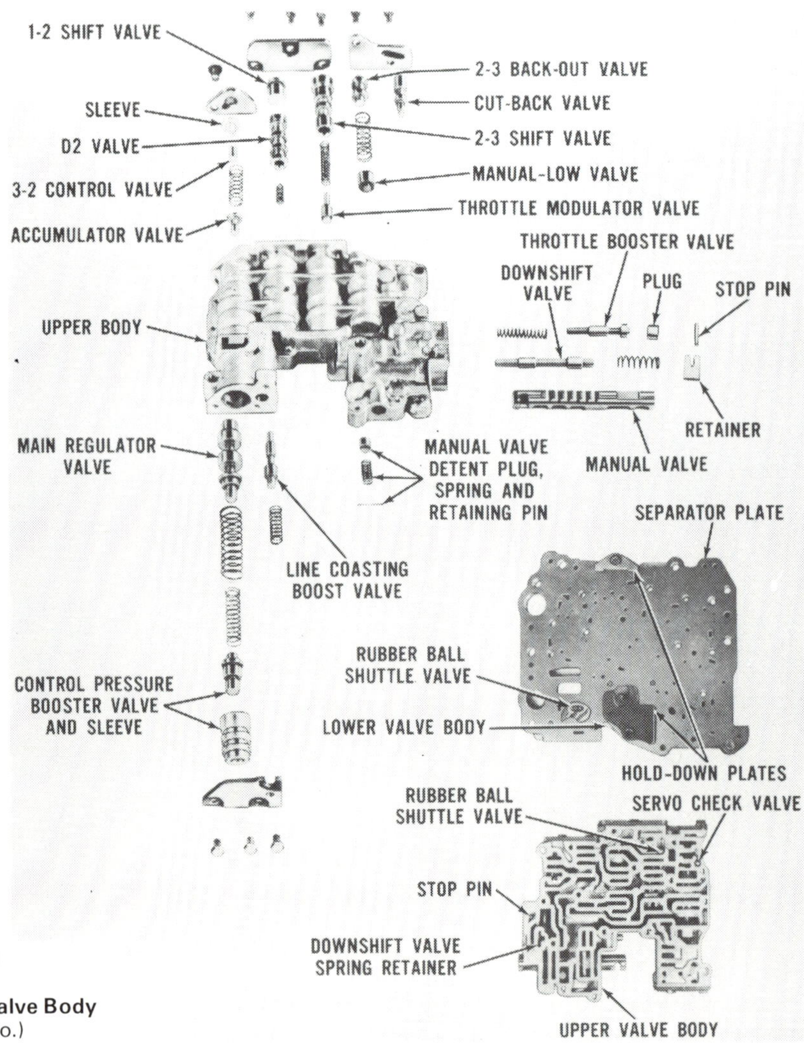

Figure 10-1

Exploded View of Valve Body
(Ford Motor Co.)

bottom of the transmission. Fluid is drawn out of the sump and through a screen or filter by the pump. The screen removes dirt and other foreign materials from the fluid and the pump then sends the fluid on to the valve body. The first valve the fluid is likely to encounter is the pressure regulator valve. All the fluid entering the valve body must pass through the pressure regulator valve which operates to maintain a specific amount of pressure for valve body operation. The type of pressure regulator usually used is a variable pressure regulator which allows different pressures for different applications. Figure 10-2 shows the path of fluid flow from the oil screen through the pressure regulator.

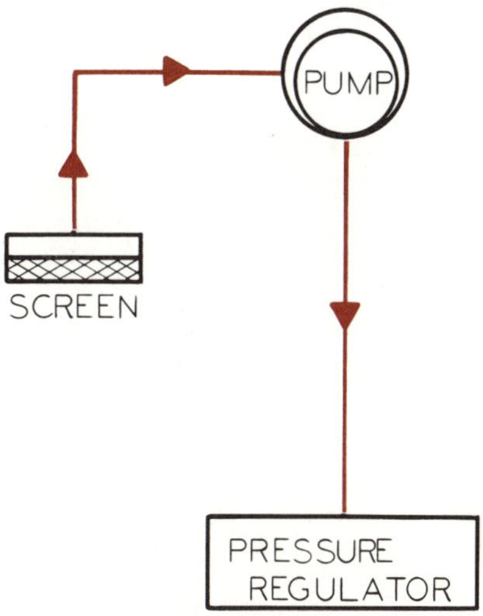

Figure 10-2

Fluid Flow from Screen to Pressure Regulator

Manual Control Valve

After the fluid passes through the pressure regulator, it enters a spool valve called the manual control valve. This valve is moved by linkage connected to the shifting quadrant inside the vehicle. The driver selects the transmission operating mode he wishes the car to be in—drive, low, neutral, park, reverse, etc.—by moving the quadrant lever into position. Movement of the quadrant linkage causes the manual control valve to be moved back and forth in its bore. The manual control valve functions to open or close passageways through which fluid enters the various circuits of the valve body.

If the driver selects a quadrant position such as low or reverse which does not require automatic shifting, the manual control valve may be the only valve in the valve body involved. Figure 10-3 shows a manual control valve connected into a valve body circuit. The fluid is pumped

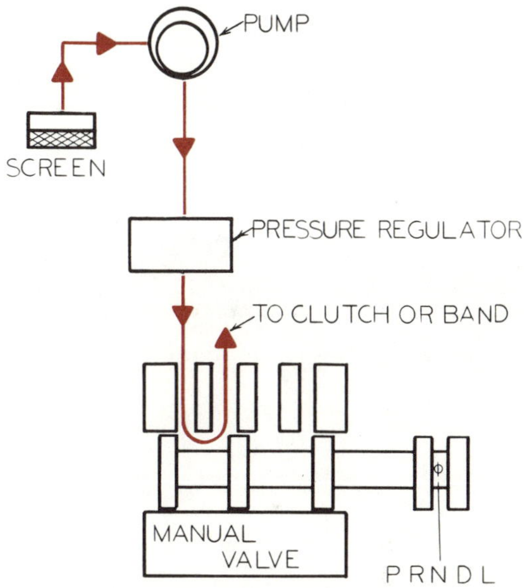

Figure 10-3

Fluid Flow through a Manual Valve

from the sump through the filter to the pressure regulator. The regulated pressure is then directed to the manual control valve. If, for example, the driver has positioned the quadrant lever in manual low, the quadrant linkage positions the manual control valve to allow fluid flow from the pump directly to the servo assembly that holds the planetary system in low gear.

Shift Valve

More valves are required to accomplish

an automatic shift than a manual shift. One valve involved in every automatic shift is the shift valve. The shift valve is a spool valve located between the manual control valve and the clutch or band servo. When the driver positions the quadrant lever in drive, the manual control valve directs fluid from the pump to a shift valve. Normally, a separate shift valve is required for each automatic shift the transmission makes. If the transmission is a three-speed, the valve body would be equipped with a 1-2 shift valve for shifting from low to second, and a 2-3 shift valve for shifting from second to high. A four-speed transmission would use 1-2, 2-3, and 3-4 shift valves.

The shift valve operates to direct fluid to the clutch or band to accomplish an upshift or downshift at the proper time. The shift valve controls two passageways or circuits, as figure 10-4 shows. When the shift valve is located in the position shown, fluid is directed to the band or clutch which holds the planetary gears in

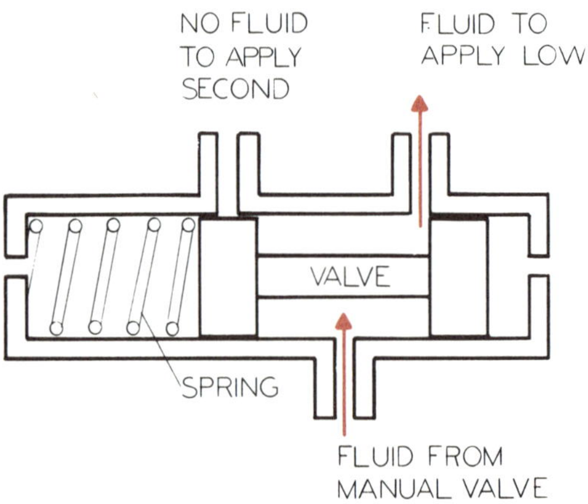

Figure 10-4

Shift Valve Directing Flow

low. When the valve is moved to the left, the low circuit is closed off. The clutches or bands responsible for low are released, because there is no longer pressure to hold them in the applied position. Fluid is now directed into the other circuit to the clutches or bands which hold the planetary system in second gear.

Figure 10-5 shows the shift valve's position in relation to the other valves in the valve body. This transmission is a three-speed, so two shift valves are required. The two shift valves are connected hydraulically so that when one valve has finished its job, the other can take over. These shift valves are positioned automatically to cause an upshift or downshift. Moving the shift valves into position automatically is the job of the governor and throttle valves.

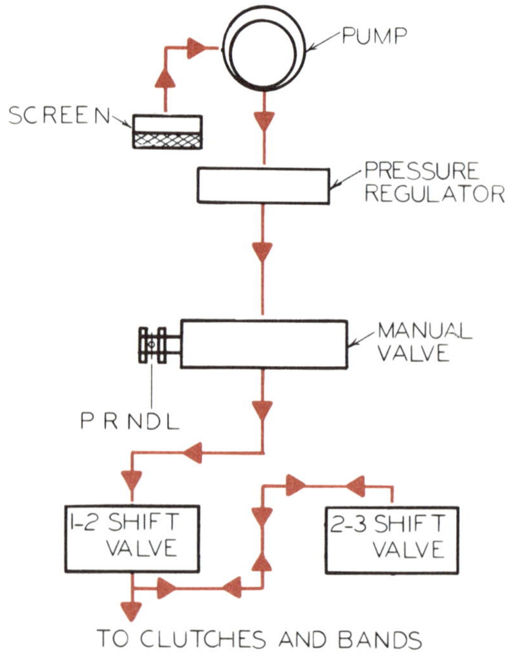

Figure 10-5

Valve Body Circuit with Two Shift Valves

Governor Valve

The governor valve is a device used to sense vehicle speed. One of the things the valve body must take into consideration for automatic shifting is how fast the vehicle is moving. The gearing requirements for the vehicle are directly related to how fast the vehicle is traveling. When the vehicle is moving away from a stop, it requires a gear advantage. As the vehicle picks up speed, it needs less of a gear advantage. If the vehicle encounters a steep hill it requires a gear advantage again. The valve body can sense this, because the vehicle slows down which affects the governor valve.

A governor valve is shown in figure 10-6. It consists of a spool valve attached

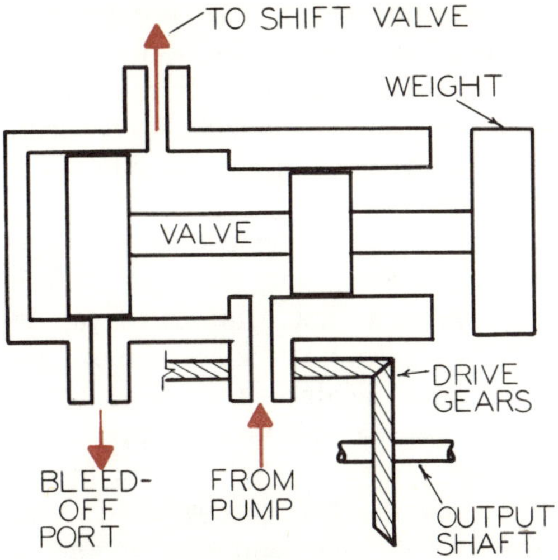

Figure 10-6

Governor Valve

to a weight and a housing which is allowed to rotate with the output shaft of the transmission. The output shaft is connected to the vehicle's drive shaft. Thus, output shaft speed is directly related to vehicle speed. As the output shaft rotates, the governor housing attached to it spins around. This spinning causes a centrifugal force to act on the governor weight, and the weight is thrown outward. The weight acts much the same as the spring in the pressure regulator valve. When the pressure regulator spring is strong the pressure is high. When the spring is weak the pressure is low. The governor weight acts the same way. When the vehicle is going slowly, the weight is not pulling hard on the spool valve. Fluid entering the governor from the pump pushes on the large land of the spool valve. The spool valve is moved to the left, and a bleed-off port is uncovered through which some of the fluid escapes, reducing the pressure. Fluid leaving the governor, called governor oil pressure, is under low pressure at this stage.

As the vehicle picks up speed, the weight is thrown out with more force. Fluid entering the governor tries to push the spool valve to the left. The fluid cannot move it as easily now, and less of the bleed-off port will be uncovered. The governor oil pressure under these conditions would be high. Thus, the governor sends along a pressure that is directly related to vehicle speed. When speed is low, governor oil pressure is low. When the vehicle is moving fast, governor oil pressure is high. This pressure signal tells the valve body how fast the vehicle is moving.

The position of the governor in relation to the rest of the valve body is shown in figure 10-7. Fluid is pumped through the pressure regulator to the governor. The governor oil pressure is routed to each of the shift valves. Governor oil pressure works to move the shift valve into a position to cause an upshift. The faster the vehicle moves, the stronger the

Figure 10-7

Governor Added to Valve Body

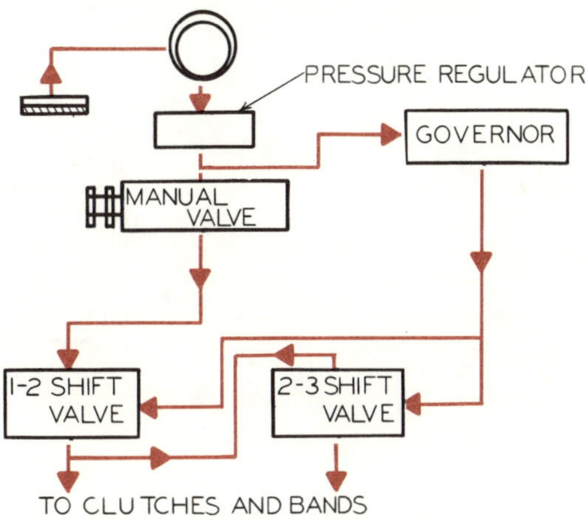

governor's push on the shift valve. An opposing force on the shift valve is provided by another valve called the throttle valve.

Throttle Valve

Not only must the valve body sense how fast the vehicle is moving, it must also sense how hard the engine is working. The vehicle might be able to breeze down the highway at 45 mph in high gear with hardly any effort at all. If the vehicle were to climb a mountain at 45 mph, however, the engine would have to work much harder. In this instance, high gear might not be appropriate. Clearly, sensing vehicle speed alone is not all that is required for automatic shifting. The device used to sense how hard the engine is working is the throttle valve.

When a driver accelerates from a stop sign with the accelerator pedal just slightly depressed, the automatic transmission shifts at a fairly low speed—25 mph, for example. On the other hand, if the driver pushes the accelerator pedal closer to the floor, the transmission does not shift until a much higher speed—45 mph, for example—is reached. Sensing how hard the engine is working, and changing the shift points accordingly, are accomplished by a throttle valve which is connected to the driver's accelerator pedal.

Figure 10-8 shows a throttle valve. The throttle valve, often called the TV, is connected at one end to some linkage which protrudes through the side of the transmission. This linkage is hooked to the accelerator pedal linkage. The other end of the valve pushes on a spring which bears on the shift valve. When the accelerator is pushed down, the throttle valve is moved in such a way as to compress the spring. Governor pressure on the other side of the shift valve is working to move the valve to the left. To do so, it must overcome the spring pressure on the throttle valve side of the shift valve.

If the driver does not have the accelerator depressed very far, the throttle valve spring is not compressed tightly. Governor pressure is able to move the shift valve against this low spring pressure at a fairly low speed (low governor pressure). If the driver pushes harder on the accelerator, the throttle valve spring is compressed, and a stronger governor pressure is required to move the shift valve. This stronger pressure can only result from a higher vehicle speed. Thus, the throttle valve is able to prevent the vehicle from upshifting until it reaches a higher speed.

Vacuum Modulator Assembly

The vacuum modulator assembly has replaced the throttle valve on most late-model automatic transmissions. The

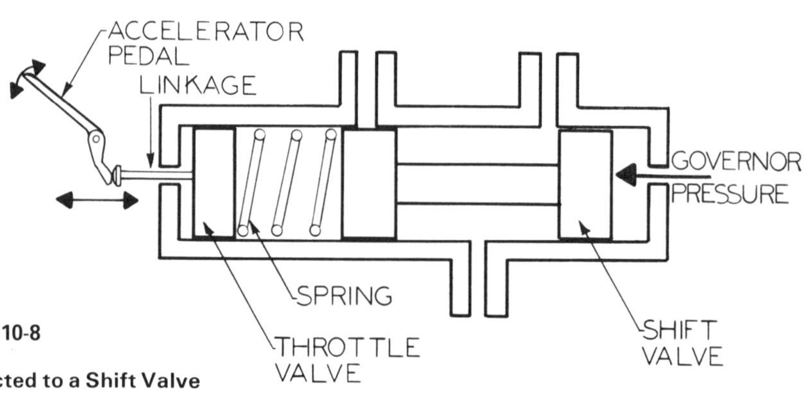

Figure 10-8

Throttle Valve Connected to a Shift Valve

vacuum modulator assembly does the same thing as the throttle valve—senses how hard the engine is working—but in a different way. The modulator directs a variable pressure to the shift valve to oppose governor pressure. This variable pressure is called modulator pressure. Modulator pressure is regulated by engine vacuum, which is an indicator of how hard the engine is working. When the vehicle is cruising down the highway with the accelerator pedal slightly depressed, the engine is not working very hard. Its vacuum under these conditions is high. If the accelerator pedal is depressed hard to climb a hill or to accelerate, the engine vacuum drops. These changes in vacuum are used to change the transmission shift points.

A vacuum modulator is shown in figure 10-9. The modulator is a metal canister mounted on the side of the transmission case. A vacuum line from the intake manifold is connected into the canister.

A sectioned view of a vacuum modulator assembly is shown in figure 10-10. The modulator assembly consists of a diaphragm stretched across a metal canister. A spring is positioned to apply a force on the diaphragm. A push rod is used to relay diaphragm movement to a valve inside the transmission. This valve may be called a throttle valve or a modulator valve. Engine vacuum acts on the diaphragm to move it against spring

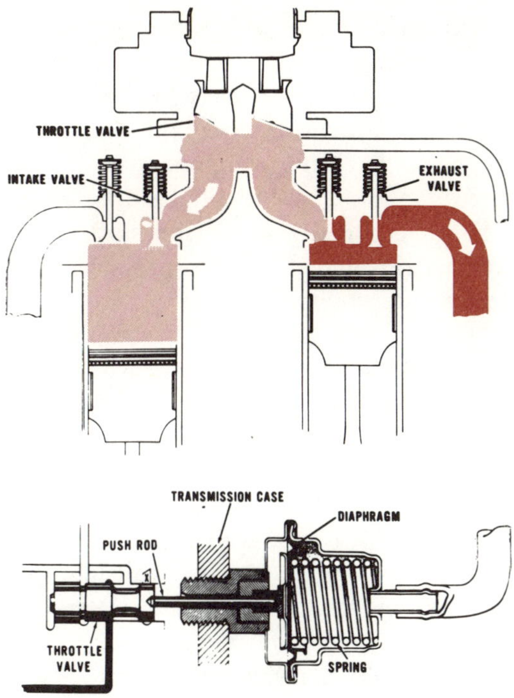

Figure 10-10

Vacuum Modulator Assembly
(Ford Motor Company)

pressure. During conditions of high engine vacuum, the diaphragm and push rod are positioned to allow the throttle valve to regulate a low modulator pressure to the shift valves. During conditions of low engine vacuum (engine working hard) the spring pushes the diaphragm and push rod in the opposite direction and causes the throttle valve to regulate a high pressure. The result is that low engine vacuum (engine working

Figure 10-9

Vacuum Modulator Mounted on Transmission

hard) equals a high modulator pressure. The governor must develop a high pressure (fast vehicle speed) to move the shift valves against modulator pressure. High engine vacuum (engine not working hard) equals low modulator pressure. The governor can move the shift valve against this low modulator pressure at a fairly low vehicle speed.

The vacuum modulator assembly has certain advantages over the throttle valve. As the vehicle begins to get old, the engine loses power. A less powerful engine requires more help from the transmission. The vacuum modulator system is able to maintain a better match between the engine and the transmission as the engine gets older, because it depends on engine vacuum. As the engine wears, its vacuum begins to diminish. This loss of vacuum results in a higher modulator pressure. The result is a transmission that stays in gear longer before it upshifts. This helps the engine to get the vehicle moving.

Another benefit of a modulator system occurs when the vehicle is operated at a high altitude, where the air is thinner and the engine needs more help from the transmission. Most vacuum modulators contain an evacuated bellows. The bellows is held in position by atmospheric pressure. As the vehicle climbs to a higher altitude atmospheric pressure is less, and the bellows begins to expand. As the bellows expands, it acts on the spring to increase modulator pressure, so the transmission is held from upshifting until a higher vehicle speed is reached. A modulator connected into a valve body circulation system is shown in figure 10-11.

Downshift or Detent Valve

Like upshifts, normal downshifts (shifting from a higher gear to a lower one) are controlled by the shift valve. When the vehicle slows down, the governor pressure on the shift valve falls off. The pressure on the modulator side of the valve is able to overcome the low governor pressure and move the valve into a position to allow fluid to apply a lower gear. There are times, however, when the vehicle must be downshifted quickly for acceleration. Forced downshifts for acceleration are controlled by a valve called a detent or downshift valve.

The detent valve is connected to the accelerator pedal. The linkage on most

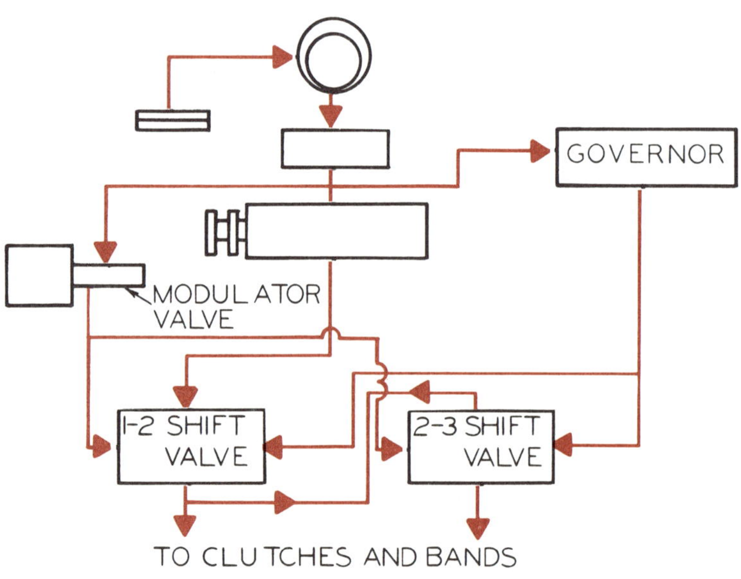

Figure 10-11

Vacuum Modulator in Valve Body Circuit

transmissions is mechanical, using a rod similar to the old throttle valve linkage connected from the accelerator pedal through the side of the transmission case into the valve body. On some transmissions the connection between the accelerator pedal and the detent is electrical. When the accelerator pedal is pushed down, a switch on the carburetor linkage is activated, and an electrical signal is transmitted to a solenoid in the transmission. The solenoid operates the detent valve.

The detent valve is a spool valve which directs pressure to the modulator side of the shift valves when it is moved. This additional pressure moves the shift valve into a downshift position against governor pressure. Governor pressure is overcome and a downshift occurs immediately. Most transmissions are protected against forced downshifts at high vehicle speeds. If the transmission were to downshift at a speed in excess of 90 mph, the transmission or engine might be damaged. The detent pressure is regulated so that it is not strong enough to oppose the governor pressure at very high vehicle speeds.

Valves that Improve Shifting

In addition to the major valves in the valve body, manufacturers may add several other valves to improve the shifting of their transmissions. Some of the more common shift improving valves are:

1. *Timing Valve.* A timing valve is used in some transmissions to make sure that two circuits are not applied at the same time. For example, in shifting from 1 to 2, the clutches or bands used for 1 must be disengaged prior to applying the clutches or bands for 2. The typical timing valve is a simple spool valve that takes a particular pressure to move. Fluid moving to apply 2 must make its way through the valve. The time it takes the fluid to move the valve and flow through it allows time for the bands or clutches used in 1 to release.

2. *Scheduling Valve.* A scheduling valve is the same as a timing valve. It times or schedules the application of one set of clutches or bands to allow time for the release of the previously used set. Other names manufacturers have used for this valve are "back-out valve" and "coasting control valve."

3. *Accumulator Valve.* In hydraulic terminology, an accumulator is a "hydraulically operated cylinder fitted with a piston, and usually a heavy spring, used to provide smooth but positive band application." An accumulator is used on some transmissions to regulate fluid flow to the servo pistons or clutches. For smooth band movement, fluid is present on both sides of a servo piston. To apply the band, fluid has to be exhausted under the servo piston as pressure is applied to it. For release, pressure is applied under the servo piston as the pressure on the top is released. The accumulator valve is responsible for directing the flow to both sides of the servo piston.

The Converter Circulation System

Fluid used in the torque converter to transmit power is continuously recirculated throughout the transmission. The valve body or pump directs the flow into and out of the converter. The circulation system is shown in figure 10-12 (next page). The circuit requires several valves. The converter pressure relief prevents the converter pressure from exceeding a safe limit. The drain-back valve prevents the fluid in the converter from leaking out through the front lubrication system when the engine is not running. The converter check valve prevents fluid in the converter from draining back through the converter return line when the engine is not running.

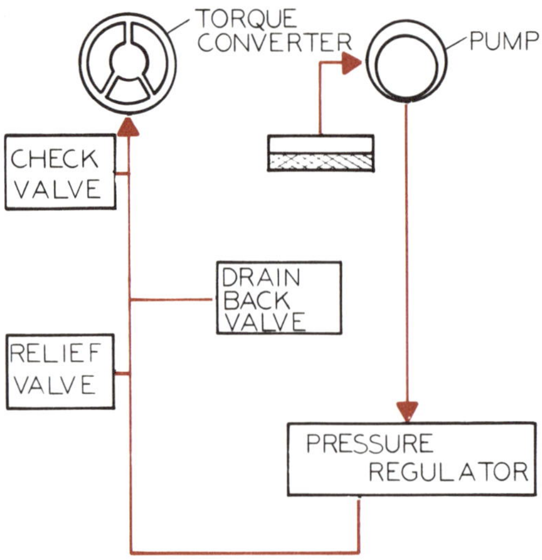

Figure 10-12

Converter Circuit

Relief Valves

Most automatic transmissions are protected against excessive pressure which might result if the normal pressure regulation system should malfunction. An unregulated pressure would immediately exceed 300 psi and cause considerable damage to many of the transmission components. An emergency relief valve is installed in the valve body circuit to relieve excessive pressure if it should occur. The valve would dump some fluid back to the sump to bleed off excessive pressure. As was already mentioned, the converter circuit also is equipped with a relief valve.

Complete Valve Body Circuit

A complete circulation diagram showing all the valves in a typical valve body is shown in figure 10-13. At first glance, the diagram may look very complex. Closer inspection, however, shows that the circuit is composed of some familiar elements—shift valves, regulator valves, governor, modulator, detent, etc.—which have been previously explained. The circulation diagram serves as a road map for the technician as he services and troubleshoots the valve body. If the technician has a basic understanding of what each valve does, he is able to follow this diagram and isolate problems. If the vehicle is having problems upshifting from 1 to 2, for example, the technician would logically begin at the 1-2 shift valve and begin tracing the circuit looking for the trouble.

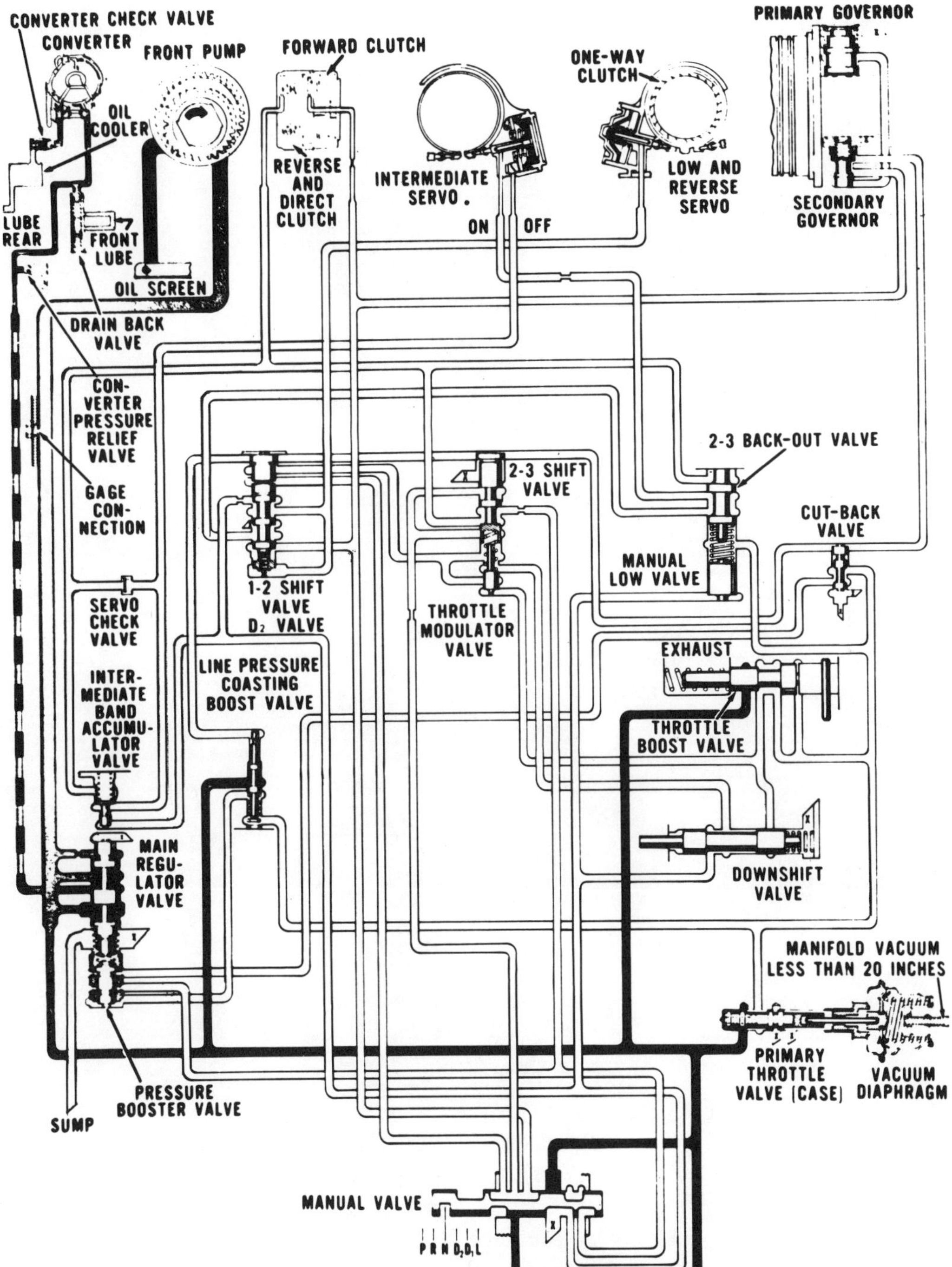

Figure 10-13
Complete Valve Body Circuit
(Ford Motor Co.)

UNIT 10
The Valve Body

True-False Questions (Indicate the correct choice)

1. The valve body is the brain of the automatic transmission. T F

2. The driver selects the transmission operating mode he wishes by positioning the manual control valve in the valve body. T F

3. The regulator valve operates to accomplish an upshift or downshift at the proper time by directing fluid to a clutch or band. T F

4. The governor valve is used to sense engine speed. T F

5. The throttle valve senses how hard the engine is working. T F

6. The vacuum modulator has replaced the throttle valve linkage on most late model transmissions. T F

7. The higher the engine vacuum, the higher the modulator pressure. T F

8. The vacuum modulator maintains a more perfect match between the engine and transmission as the vehicle gets older. T F

9. Forced downshifts for acceleration are controlled by a detent valve. T F

10. Timing valves and scheduling valves are used to improve transmission shifting. T F

Multiple Choice Questions (Indicate the correct choice)

1. The valve which is connected to the shift quadrant linkage is:
   a. governor valve
   b. shift valve
   c. modulator valve
   d. manual control valve

2. The shift valve is positioned by pressure from:
   a. governor
   b. modulator
   c. regulator
   d. both *a* and *b*

3. The valve which senses how fast the vehicle is traveling is the:
   a. governor valve
   b. manual valve
   c. throttle valve
   d. modulator valve

4. The valve which senses how hard the engine is working is the:
   a. governor valve
   b. manual valve
   c. modulator valve
   d. timing valve

5. When the vehicle pulls away from a stop sign with the throttle depressed hard, the transmission stays in low gear longer than it does when the throttle is slightly depressed. In this case, the shift valve is responding to:
   a. governor pressure
   b. modulator pressure
   c. main line pressure
   d. regulator pressure

Parts Identification

Identify the components of a complete valve body shown.

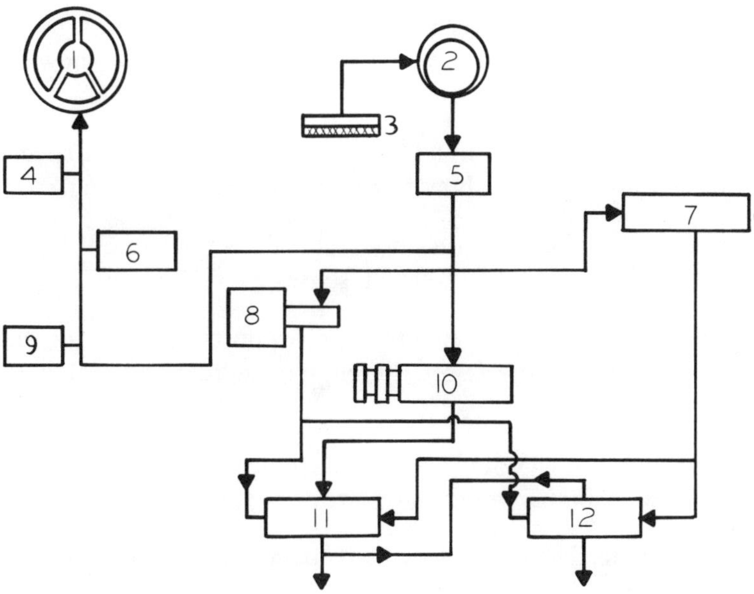

1. ____________	5. ____________	9. ____________
2. ____________	6. ____________	10. ____________
3. ____________	7. ____________	11. ____________
4. ____________	8. ____________	12. ____________

UNIT 10
The Valve Body

True-False Questions

1. This statement is true. The valve body senses the vehicle's gearing requirements and selects the correct gear ratio in the transmission.

2. This statement is true. The manual control valve is moved by linkage connected to shifting quadrant. The manual control valve functions to open or close passageways for fluid to enter the various circuits for "drive," "low," "neutral," etc.

3. This statement is false. The shift valve, not the regulator valve, is responsible for causing an upshift or downshift at the proper time by directing fluid at a clutch or band.

4. This statement is false. The governor valve is a device used to sense vehicle speed. The gearing requirements for the vehicle are directly related to how fast the vehicle is traveling.

5. This statement is true. Not only must the valve body sense vehicle speed, it must also have a throttle valve to sense how hard the engine is working.

6. This statement is true. The vacuum modulator does the same thing as the throttle valve linkage on older transmissions: it senses how hard the engine is working.

7. This statement is false. Engine vacuum operates on the modulator diaphragm and spring in such a way that the *lower* the engine vacuum the higher the modulator pressure.

8. This statement is true. As a vehicle gets older the engine vacuum drops. This loss of vacuum results in a higher modulator pressure and a change in shift points.

9. This statement is true. The detent valve directs pressure to the modulator side of the shift valve, forcing a downshift.

10. This statement is true. Timing valves and scheduling valves are used to make sure that two circuits are not applied at the same time, thus improving shifting.

Multiple Choice Questions

1. The correct choice is *d*. The manual valve is controlled by the driver through linkage connected to the shift quadrant.

2. The correct choice is *d*. The shift valve is positioned by opposing pressures developed by the governor and modulator valves.

3. The correct choice is *a*. The governor valve, connected to the transmission output shaft, senses vehicle speed.

4. The correct choice is *c.* The modulator, connected to engine vacuum, senses how hard the engine is working.

5. The correct choice is *b.* The shift valve is responding to a high modulator pressure and holding the transmission in low gear longer.

Parts Identification

The components of a complete valve body are:

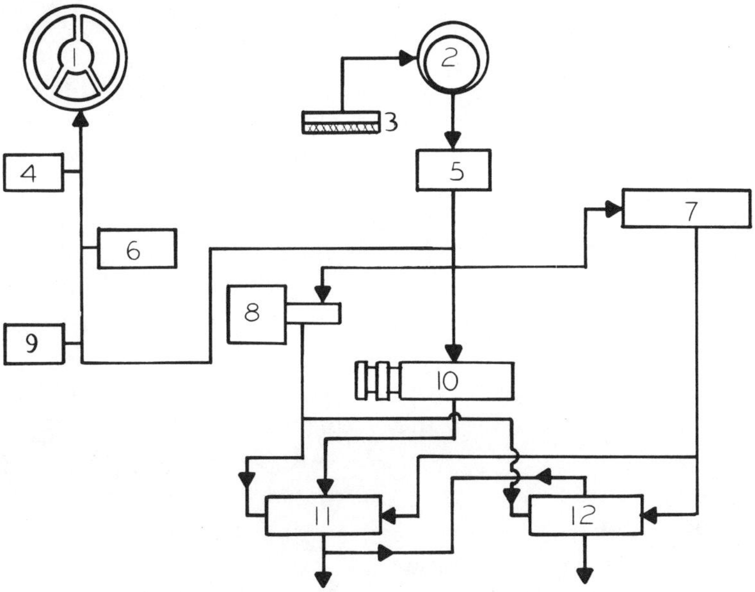

1. Torque Converter
2. Pump
3. Screen
4. Check Valve
5. Pressure Regulator
6. Drain Back Valve
7. Governor
8. Modulator
9. Relief Valve
10. Manual Valve
11. 1-2 Shift Valve
12. 2-3 Shift Valve

Automatic Transmission Cooling Systems

The automatic transmission with its torque converter brought a cooling problem not present in the hand-shifted standard transmission. Slippage in the torque converter, clutches, and bands generates a tremendous amount of heat. This heat must be removed from the transmission or serious damage might result. This unit will describe the methods used to cool automatic transmissions.

The Cooling System's Job

An automatic transmission cooling system has three important functions.

1. *The cooling system must provide adequate cooling during conditions of extreme vehicle operation and high ambient (outside) temperatures.* City driving, for example, presents one of the most difficult cooling problems because engine speeds are low and much time is spent at idle with the converter slipping. A large transmission rebuilder in New York estimated that, due to the extremely congested traffic conditions and the narrowness of streets, restricting air cooling, transmission life in his area is only 30,000 miles. Hill climbing and trailer towing are other driving conditions generating large amounts of heat.

2. *The cooling system must permit the fluid to rise to a satisfactory operating temperature in the shortest possible time after vehicle start-up.* As explained earlier, the thickness or viscosity of the fluid changes with changing temperatures. When weather conditions are very cold, the fluid is thick. If the transmission fluid is not warmed quickly, the normal shifting and lubrication of the transmission will be adversely affected.

3. *The cooling system must maintain the best possible fluid temperature for normal vehicle operation.* All transmissions do not have the same optimum temperature. Most manufacturers feel the best operating range is a temperature between 150° and 250°F. Engineers consider that adequate cooling has been provided if temperatures can be limited to about 275°F.

The Effects of Overheating

Studies have demonstrated that automatic transmissions last longer if they can be prevented from overheating. Excessive amounts of heat destroy the nonmetallic materials used in friction facings and seals. The effect of heat on transmission fluid has received a great deal of attention in recent years. It was determined that fluid fails as a result of being exposed to high temperatures for prolonged periods of time. The fluid can be considered as having failed when sludge and varnish have formed to the degree that these deposits will not be removed by draining. Temperature, too, has an effect on the transmission operation. Most noticeable to the driver is the shift "feel." The transmission shifts objectionably as temperature works on the frictional and sealing materials.

Air Cooling Systems

Air cooling of the automatic transmission has been used to some extent since the first automatic transmission was introduced. On the first transmissions, the flow of air passing normally over the outside of the transmission provided sufficient cooling. When vehicle power-to-weight ratios increased and torque converter transmissions became available, normal surface cooling was not enough. Forced air cooling was required to operate the transmission at a safe temperature.

An air-cooled automatic transmission uses the torque converter housing to circulate the air. The path of air flow is shown in figure 11-1. Fins are positioned

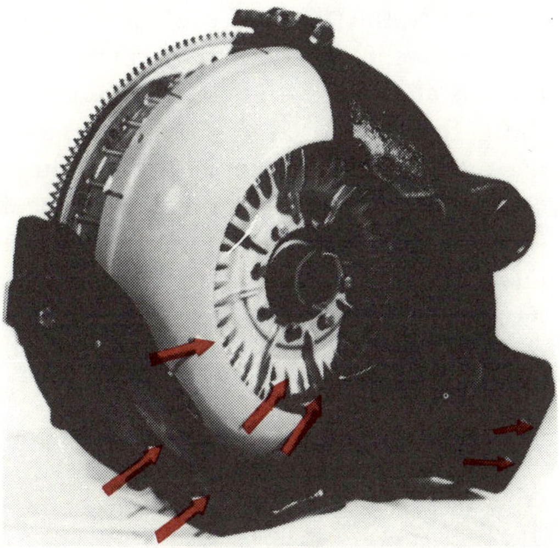

Figure 11-1

Path of Air Flow in an Air-Cooled Automatic Transmission

on the outside of the torque converter. As the converter rotates, it acts as a centrifugal air pump or fan, and pumps cool outside air around the hot outer surface of the torque converter. The heated air is then routed outside. The ducting that routes the air provides for an air inlet (usually with a screen to keep out foreign material), a shroud to guide the air over the converter, and an air outlet.

The transmission size and arrangement determine whether air cooling can be utilized. Air cooling is particularly suited to transmissions with a large converter cooling surface. Units with a small converter cooling surface cannot be cooled as efficiently by air. Air cooling is, of course, the most appropriate method of cooling a transmission mated to an air-cooled engine. Figure 11-2 shows an

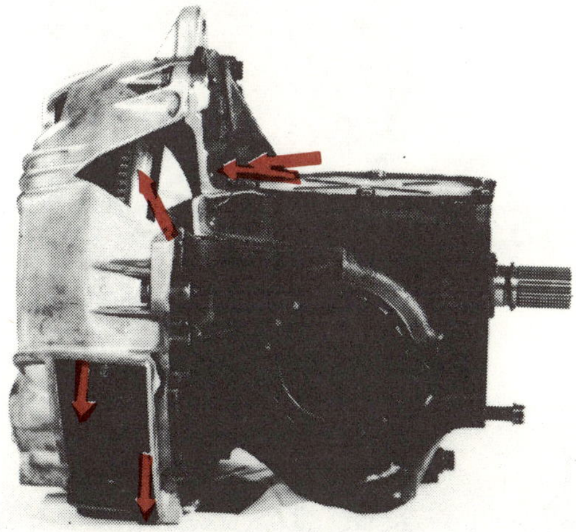

Figure 11-2

Air Cooling Used with an Air-Cooled Engine

air-cooling system used on a transaxle unit similar to those mounted on most air-cooled engines.

Liquid Cooling Systems

The most common method of cooling automatic transmissions is liquid cooling. The liquid cooling system utilizes a heat exchanger located in the bottom tank of the engine's radiator. A heat exchanger is a device used to exchange heat from one medium to another—in this case, from the fluid to the cooling water. The heat exchanger is constructed from a metal such as copper, with good heat conducting ability. There are many passages in the heat exchanger for the fluid to flow through. This arrangement allows the fluid to contact the largest possible metal surface area. The heat is transmitted through the metal to the cooling liquid. A typical heat exchanger installation is shown in figure 11-3. Hot fluid is pumped from the transmission by the front pump through the heat exchanger and then back into the transmission.

Figure 11-3

Heat Exchanger Located in Bottom Tank of Radiator

A typical circulation diagram for a liquid cooling system is shown in figure 11-4. Fluid is moved from the pump through the pressure regulator. The cooling circuit is equipped with a by-pass orifice (or some similar device) to lower the pressure below main line pressure. This allows a lower pressure to be used in the lines which leave the transmission, and a lower pressure also in the cooler. This lower pressure is less likely to cause leaks which would result in a serious loss of fluid. From the pressure regulator the fluid is routed through the torque converter. Fluid that is heated in the torque converter is routed outside the transmission to the heat exchanger and back into the transmission.

Auxiliary Cooling Systems

When vehicles are used in extreme temperatures or under severe operating conditions, the standard cooling system is not sufficient. Vehicles that spend long periods at idle, such as police vehicles or those that are used for trailer towing, need more cooling capacity. Auxiliary coolers have been developed to meet these conditions. An auxiliary cooler is a large heat exchanger that can be mounted next to the vehicle's radiator in the air stream. These heat exchangers may be installed in place of, or in addition to, the conventional heat ex-

Figure 11-4

Circulation Diagram for Water Cooling

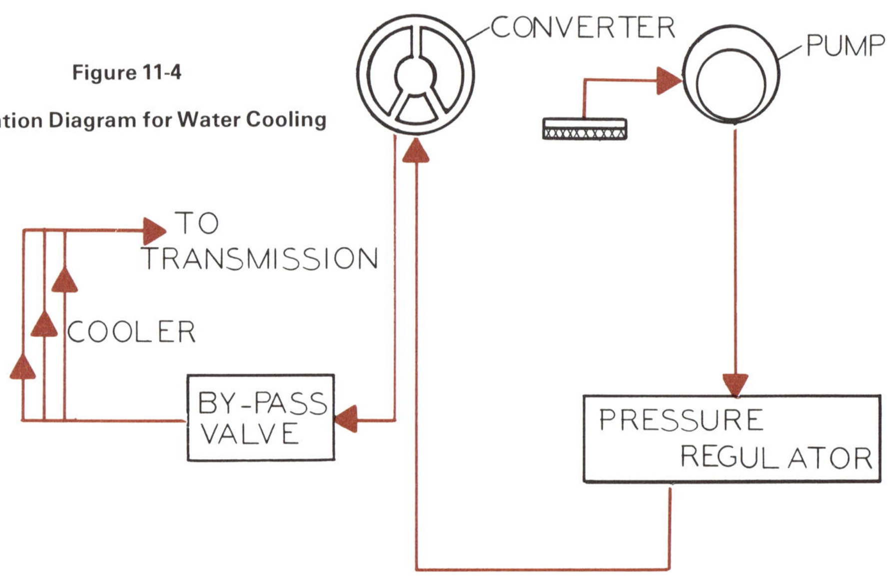

changer. The auxiliary cooler may be supplied by the vehicle manufacturer as optional equipment on trailer towing packages. It is also supplied by independent manufacturers to be installed as an "after market" item. An auxiliary cooling system is shown in figure 11-5.

The installation of an auxiliary cooling system is a relatively simple operation requiring an electric drill, a tubing cutter or hacksaw, a screwdriver, and a combination wrench. The heat exchanger is mounted in front of the radiator in the cooling air stream as shown in figure

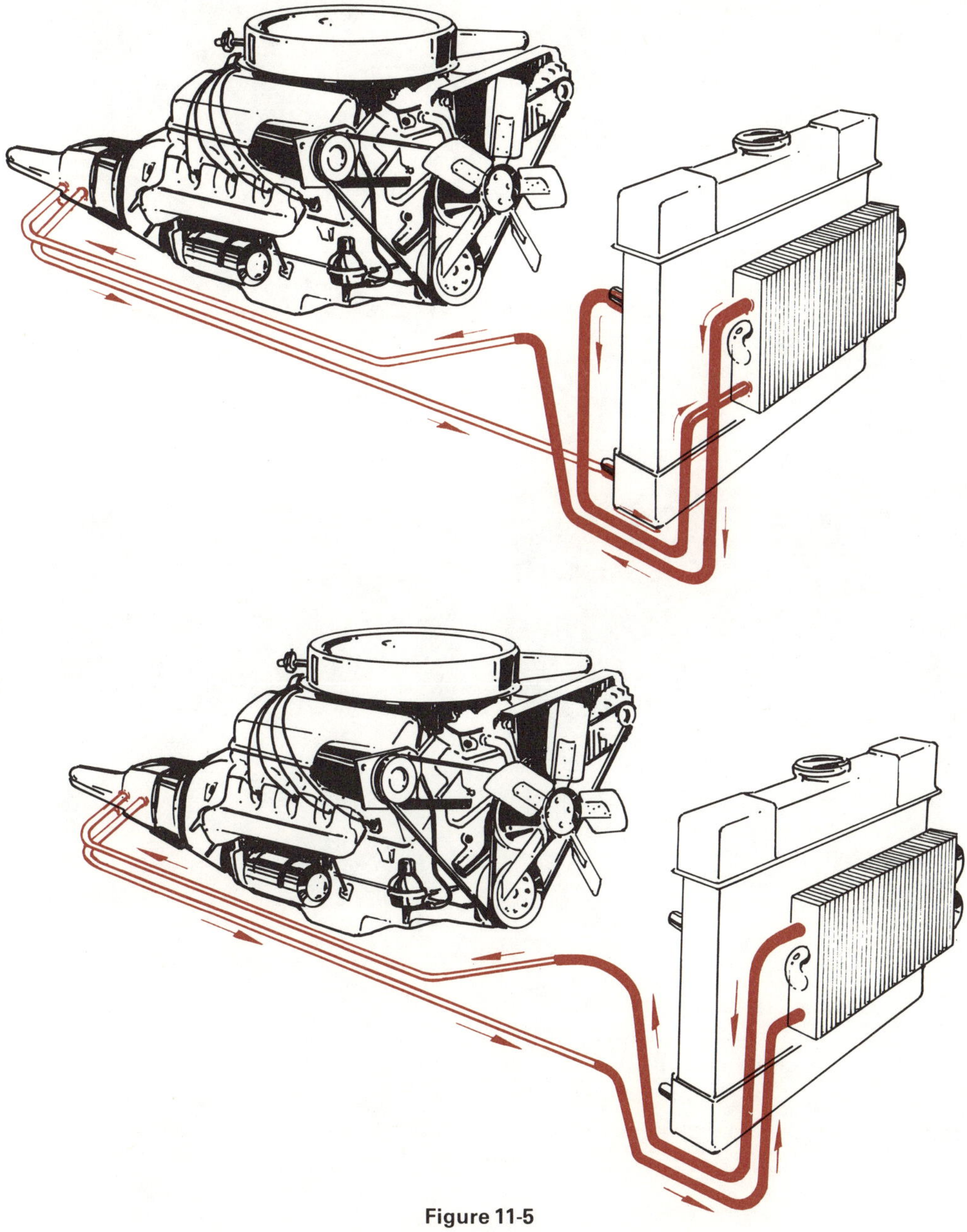

Figure 11-5

An Auxiliary Cooler May Be Installed in Series with the Radiator Heat Exchanger (top) **or Used Separately** (bottom)
(*Trailer Life*, April 1971)

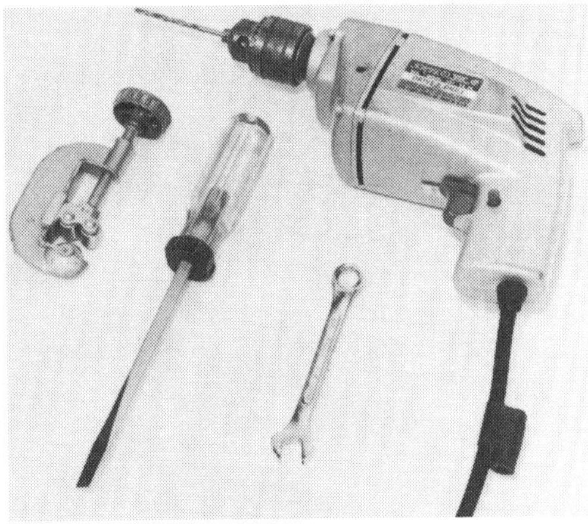

Figure 11-6

Tools Required to Install an Auxiliary Cooler
(Hayden Trans-Cooler, Inc.)

Figure 11-8

Cutting the Fluid Return Line
(Hayden Trans-Cooler, Inc.)

11-7. The transmission fluid return line is cut near its connection to the radiator. Hoses are then connected from the heat exchanger to the return line to place the auxiliary heat exchanger in series with the factory heat exchanger in the radiator bottom tank.

Figure 11-7

Mounting the Heat Exchanger in Front of the Radiator
(Hayden Trans-Cooler, Inc.)

Figure 11-9

Installing Hoses on the Return Line
(Hayden Trans-Cooler, Inc.)

Proper installation of the auxiliary cooling system is important to maintain car warranty. The manufacturer's recommendations should be followed.

UNIT 11
Automatic Transmission Cooling Systems

True-False Questions (Indicate the correct choice)

1. City driving and trailer towing are both conditions that present difficult cooling problems for an automatic transmission. T F

2. Most manufacturers feel that the best operating range for an automatic transmission is a temperature between 150° and 250° F. T F

3. The cooling system must operate to maintain the lowest possible fluid temperature. T F

4. Automatic transmissions last longer if they can be prevented from overheating. T F

5. Air cooling has been used to some extent since the first automatic transmission was introduced. T F

6. The most common method of cooling an automatic transmission is forced air circulation. T F

7. The device used to remove the heat from an automatic transmission is called a heat exchanger. T F

8. Air cooling would be most appropriate for an air-cooled vehicle. T F

9. With today's cooling systems, even transmissions operating under severe temperature conditions are properly protected from overheating. T F

UNIT 11
Automatic Transmission Cooling Systems

True-False Questions

1. This statement is true. City driving and trailer towing both present difficult cooling problems for an automatic transmission.

2. This statement is true. Manufacturers prefer a temperature range between 150° and 250° F. Engineers consider that adequate cooling has been provided if temperatures can be limited to about 275° F.

3. This statement is false. The cooling system must permit the fluid to rise to a satisfactory operating temperature in the shortest possible time after vehicle start-up. If the fluid is too cold, normal shifting and lubrication will be adversely affected.

4. This statement is true. Excessive amounts of heat destroy the non-metallic materials used in friction facings and seals.

5. This statement is true. On the first transmissions the flow of air passing normally over the outside of the transmission provided sufficient cooling.

6. This statement is false. Water cooling, not forced air circulation, is the most common method of cooling an automatic transmission.

7. This statement is true. A heat exchanger, located in the bottom tank of the radiator, removes the heat from the fluid and passes it into the cooling water.

8. This statement is true. Air cooling is particularly suited to transmissions that are mated to air-cooled engines.

9. This statement is false. When vehicles are used in extreme temperatures or under severe operating conditions, the standard cooling system is not sufficient for proper protection from overheating. Auxiliary coolers have been developed to meet these conditions.

Section V

MAINTENANCE AND SERVICE

Automatic Transmission Maintenance

If the owner of a vehicle expects long service from his automatic transmission he must maintain it properly. There are several preventative maintenance operations that should be performed periodically on an automatic transmission to keep it operating in top condition. This unit outlines these procedures.

Fluid Level Check

The level of fluid in an automatic transmission is critical to its operation. If the fluid level is too high or too low there may be a disruption in the normal shifting and a loss of lubrication. The fluid level should be checked at frequent intervals. A safe rule would be to check the fluid level every second or third gasoline fill-up.

When the fluid is checked, several precautions should be observed. The engine and transmission must be at normal operating temperature. The fluid expands as it warms and a fluid level check on a cold transmission may show a low level. The transmission must be in the correct quadrant position for that vehicle. Some transmission fluid levels are checked in "park," others in "drive," and so on. If the level is checked with the transmission in the wrong operating mode, the level may appear to be too high or too low when it is actually correct. This is because the fluid level in the sump will vary as the transmission goes into different operating modes. In some modes, clutch drums and servos are filled with fluid, causing a low level in the sump. In other modes, all the fluid is exhausted from servos and clutches, causing a high fluid level. The correct operating mode for checking the fluid level is found in the vehicle owner's manual or in a transmission service manual.

The procedure for checking the fluid level is as follows:

1. Drive the vehicle onto a level floor.
2. Set the emergency brake firmly.
3. Start the engine and let it idle.
4. Move the quadrant lever through each of the operating modes and then position it in the correct mode for checking the fluid level.
5. Remove the transmission dip stick and wipe it clean.
6. Push the dip stick back into position firmly.
7. Remove the dip stick and read the level. Dip sticks normally have two marks; one for *full* and one for *low* or *add*. The low mark on many dip sticks represents a pint. There is, however, *no standard marking system for dip sticks. Check the vehicle owner's manual for proper interpretation.*
8. Add the proper amount of ATF, as required. *Be sure to use the correct type of fluid.* (See Unit 2).
9. Replace the dip stick.

Fluid Change

Most manufacturers recommend that automatic transmission fluid be changed at regular intervals. The interval specified by the manufacturer is based on a time or mileage interval. These recommendations can be found in the vehicle owner's manual. In the absence of such recommendations, the following intervals could be used. For mild service conditions—driving that does not generate high temperatures—fluid should be changed every 20,000 to 25,000 miles. For severe service—such as heavy traffic with long periods at idle or trailer towing—the interval should be reduced to 6,000 to 10,000 miles.

A fluid change involves more than just draining and refilling the sump. Every fluid change should include the removal of the pan and cleaning or replacement of the filter screen. A typical fluid change procedure follows:

1. Jack or hoist vehicle off the ground.
2. Inspect fluid pan for drain plug—most late-model vehicles *do not* have a drain plug. This insures pan removal during a fluid change.
3. If the pan has a drain plug, remove it and allow the fluid to drain. Then remove the pan. (Caution: It may be hot!)

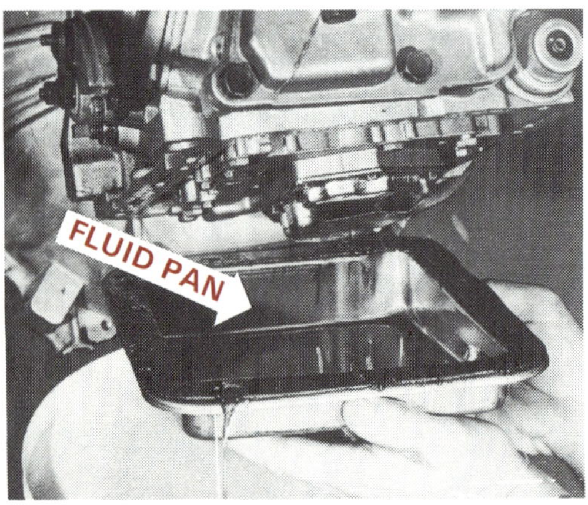

Figure 12-1

Fluid Pan Being Removed

4. On units without a drain plug, remove pan to drain fluid.
5. Remove flywheel inspection plate.
6. Rotate engine and check for drain plug on converter or fluid coupling. Many late-model converters do not have plugs and cannot be drained.

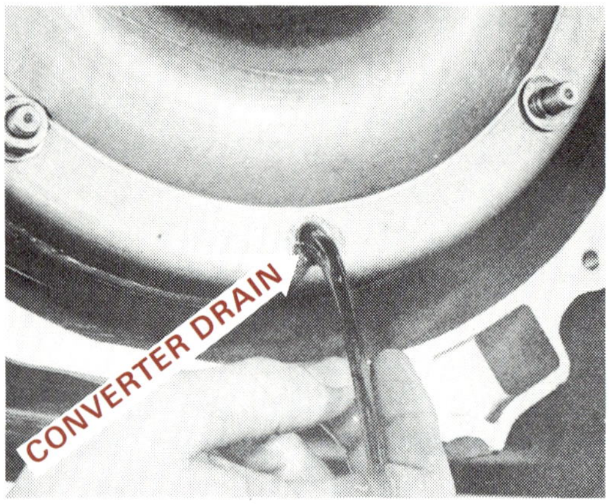

Figure 12-2

Converter Drain Plug

7. If the converter can be drained, it will have two plugs—one for a vent and one for a drain. Remove both plugs and drain the converter. Replace the plugs.
8. Remove filter or filter screen.

Figure 12-3

Removing the Filter Screen

9. Clean or replace filter as specified by the manufacturer.
10. Clean the fluid pan.
11. Install pan with new gasket, using torque wrench to tighten cap screws.

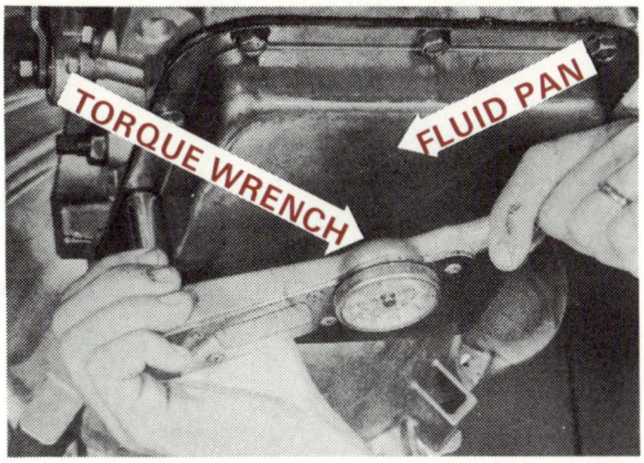

Figure 12-4

Installing the Fluid Pan (After installation the bolts are torqued with a torque wrench.)

12. Check ATF capacity of transmission in owner's manual. If converter was drained, return only a portion of the fluid into the transmission.

13. Start engine and allow time for warm-up.

14. Check fluid level.

15. Add additional fluid as required to bring the unit to the correct level.

Adjustments

Several periodic adjustments are required on most automatic transmissions. To perform these adjustments correctly, the technician must follow carefully the procedures outlined in a transmission service manual. The following adjustments are typical:

1. *Band Adjustment.* In order to function correctly, an automatic transmission's bands must be in proper adjustment (correct amount of space between drum and frictional material on the band). A correctly adjusted band provides a good positive hold on application, will not drag on release, and provides smooth engagement with correct timing. Older transmissions required a band adjustment at frequent intervals to account for wear on the bands. New frictional materials used on late-model transmissions nearly eliminated the need for this adjustment. When a band began to slip on older transmissions, a band adjustment would usually solve the problem. A slipping band on a late-model unit usually indicates a loss of pressure for band application and cannot be cured by an adjustment.

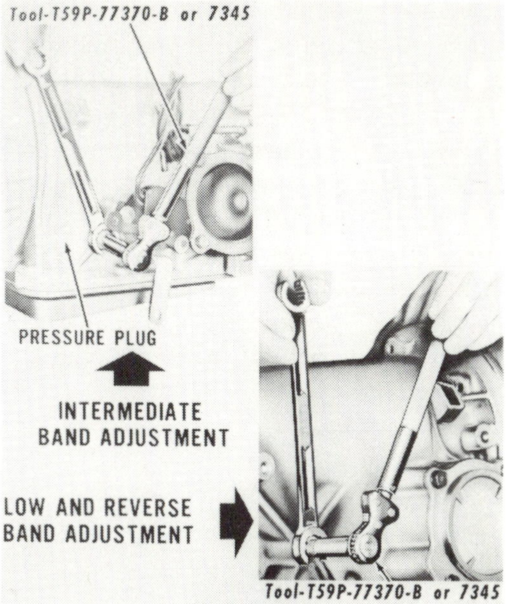

Figure 12-5

Adjusting a Band
(Ford Motor Co.)

2. *Manual Control Linkage.* The linkage from the shift quadrant to the manual control valve is called the manual control linkage. This linkage is adjustable. If the linkage is not in correct adjustment, the shift quadrant will not

Figure 12-6

Manual Control Linkage

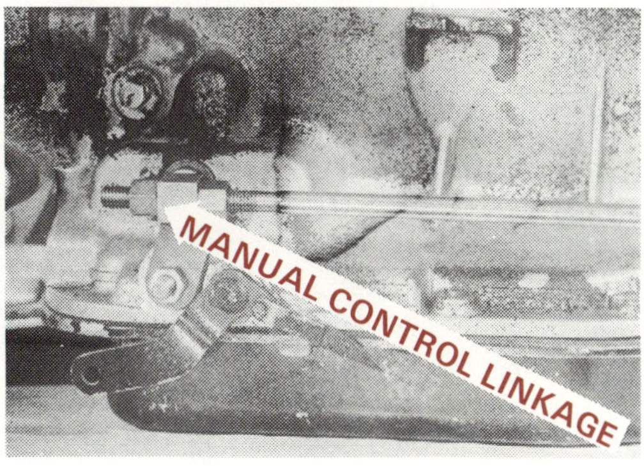

position the manual control valve correctly which may cause cross feed. Cross feed means that two circuits are pressurized when only one should be. The result is a transmission that cannot decide what mode to be in. Normally the manual control linkage would not require periodic adjustment, but it may require adjustment if the linkage is disturbed.

3. *Throttle Linkage.* The linkage from the throttle to the automatic transmission detent valve or throttle valve is also adjustable. This linkage would not normally require periodic attention but might require attention if it is disturbed. Incorrect adjustment of throttle linkage may cause a variety of problems related to shift timing. If the transmission shifts too early, or too late, or will not downshift, the throttle linkage may be at fault.

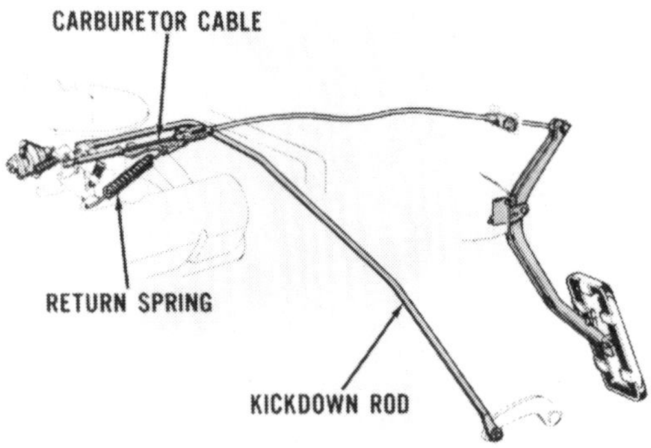

Figure 12-7

Throttle Linkage
(Ford Motor Co.)

UNIT 12
Automatic Transmission Maintenance

True-False Questions (Indicate the correct choice)

1. The transmission must be warm when the fluid level is checked. T F

2. All automatic transmissions must be in "drive" for a fluid level check. T F

3. The "add" mark on a transmission dip stick indicates the fluid level is one pint low. T F

4. The oil change interval for mild service conditions is 20,000 to 25,000 miles. T F

5. The oil change interval for severe service conditions is 6,000 to 10,000 miles. T F

6. A fluid change only involves draining and refilling the sump. T F

7. Most late-model automatic transmissions have a drain plug in the pan for draining the fluid. T F

8. Late-model automatic transmissions require frequent band adjustments. T F

9. A slipping band on a late-model automatic transmission usually indicates a need for a band adjustment. T F

10. Manual control linkage and throttle linkage require periodic adjustment. T F

UNIT 12
Automatic Transmission Maintenance

True-False Questions

1. This statement is true. The fluid expands as it warms, and a fluid check on a cold transmission may show a low level.

2. This statement is false. Some transmission fluid levels are checked in "drive," others in "park," and so on.

3. This statement is false. There is no standard marking system for dip sticks.

4. This statement is true. For mild service conditions—driving that does not generate high temperatures—fluid should be changed every 20,000 to 25,000 miles.

5. This statement is true. For severe service conditions such as heavy traffic or trailer towing, the fluid should be changed every 6,000 to 10,000 miles.

6. This statement is false. Every fluid change should include the removal of the pan and a cleaning or replacement of the filter screen.

7. This statement is false. Most late-model transmissions do not have a drain plug. This insures pan removal during a fluid change.

8. This statement is false. New frictional materials used on late-model transmissions have nearly eliminated the need for a band adjustment.

9. This statement is false. A slipping band on a late-model transmission usually indicates a loss of pressure for band application.

10. This statement is false. Adjustments on the manual control and throttle or detent valve linkages are required only if linkages have been disturbed. They do not require periodic adjustment.

Automatic Transmission Troubleshooting

Accurate diagnosis of apparent transmission problems begins with a thorough understanding of how a transmission works. Armed with this knowledge and appropriate service publications, the technician can isolate and cure automatic transmission problems. It should be pointed out that there is no substitute for the transmission manufacturer's trouble-shooting guide. The technician must use the diagnostic procedure outlined by the manufacturer of the transmission. The purpose of this unit is to explain the general procedures used in troubleshooting, and to point out the importance of a logical diagnostic sequence.

Preliminary Checks

When the technician encounters a transmission that is suspected of malfunctioning, he must perform a step-by-step diagnosis to isolate the cause and effect a cure. Troubleshooting an automatic transmission is a three-phase operation. First, the technician must perform some preliminary checks to eliminate any obvious problems. Second, he should road test the vehicle to experience the problem or complaint for himself. Finally, he should perform a shop diagnosis. The preliminary checks to be made are:

1. *Engine Condition.* The best operating transmission in the world will not make the vehicle perform if the engine cannot deliver the power. The technician should start the engine and see that it starts quickly, runs smoothly, and accelerates without smoking. If there is any reason to suspect that the engine is out of tune or in need of major service, this should be done prior to further diagnosis of transmission problems.

2. *Fluid Level.* As mentioned earlier, the fluid level is critical to the operation of the automatic transmission. Both the level and condition of the fluid can provide valuable clues to the general condition of the transmission. A low fluid level would allow the pump to take in air along with the fluid. Air in the hydraulic system makes the fluid spongy and compressible. Pressure builds up too slowly, resulting in delayed shifts and slipping. The slipping causes overheating and rapid wear of clutches and bands. If the fluid level is very low it will starve the pump to the extent that adequate lubrication cannot be maintained, and other transmission components will also be damaged.

If the fluid level is too high, other problems occur. The gears in the transmission churn like an eggbeater, causing the fluid to foam. The air bubbles in the foam cause the same slipping and overheating conditions caused by low fluid level. Even worse, the heat and foam cause the fluid to oxidize. As was pointed out earlier, oxidizing the fluid causes it to form a varnish. A varnish build-up can quickly impair valve operation.

3. *Fluid Condition.* The condition of the fluid is as important as the level. When the fluid level is checked, the tech-

nician should closely inspect the fluid on the dip stick. If the fluid is discolored or shows signs of foreign material, the transmission may need major service. If the fluid is very dark and smells burned, frictional material has been overheated. A complete overhaul will probably be necessary to replace the lost frictional material.

Fluid which has a foamy, milky appearance indicates that water has entered the transmission. Water may enter the transmission through a broken heat exchanger in the radiator. Another possibility is that rain water may have entered through a raised dip stick sealing cap. Water mixed with transmission fluid swells transmission seals and softens frictional material.

A varnish build-up in the transmission is evidenced by a dip stick that is tacky and will not wipe clean. Once varnish starts forming, it builds up in all the valves, servos, and clutches, and causes sticking. Sticking valves may cause erratic transmission operation. Eventually the varnish formation will clog the filter or screen. Pump pressure will drop and the transmission may fail to operate at all.

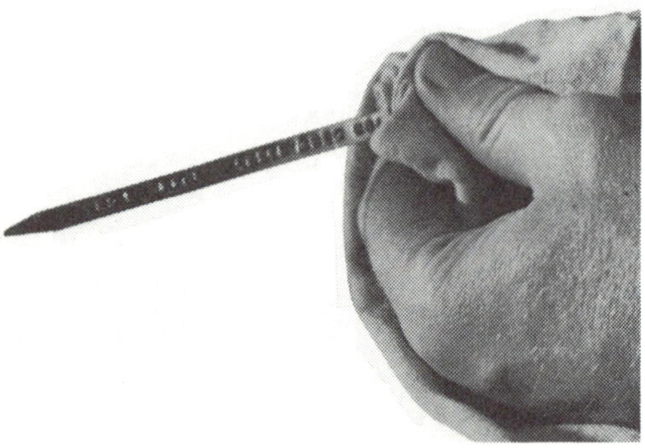

Figure 13-1

Varnish Build-Up Is Evidenced by a Dip Stick that Will Not Wipe Clean

4. *Control Linkage.* Another preliminary check is the manual and throttle control linkage. Many of the complaints on improper transmission operation can be eliminated by a correct setting of the control linkage. One of the signs of improper manual control linkage setting is a failure of the engine to start unless the quadrant lever is moved slightly off "neutral" or "park" detents. Another indication is creeping when the selector is in "neutral." Improper throttle linkage setting results in an improper shift feel. The shift feels harsh or slippery. The technician should inspect all the linkage for signs of damage or distortion.

5. *Vacuum Modulator.* The last preliminary check is a test of vacuum modulator operation. A transmission that is having trouble shifting can be suspected of vacuum modulator trouble. To check the vacuum modulator, start the engine and allow it to idle. Pull the vacuum hose off the vacuum modulator

Figure 13-2

Checking the Vacuum Modulator by Removing the Hose

canister. The engine speed should change significantly as the vacuum is lost through the open hose. If engine speed does not change, the vacuum hose should be checked for obstructions and leaks. If the hose is in good shape, a faulty diaphragm in the modulator may be the trouble.

Road Test

If preliminary tests fail to uncover a problem, the vehicle should be road tested. If possible, a standard test route should be developed so that a proper comparison can be made between vehicles. The car's owner should accompany the technician and his complaint be analyzed as the problem occurs. The test route should include a level section, some hills, and a quiet section for checking noises.

On a road test, the transmission should be operated in each of its ranges to determine how the shifting feels, and to check upshift and downshift speeds. The most important thing to look for is evidence of slipping. Slipping in any gear means band or clutch trouble. If only one band or clutch is at fault it may be isolated for repair. The technician should make a note of the operating mode where the transmission slips. He can then determine which band or clutch is causing the problem by referring to a band and clutch application chart. These charts show what band or clutch is applied in each operating mode. The technician can then isolate the problem to a particular band or clutch for repair.

A large percentage of troubleshooting time will be involved with noise diagnosis. Accurate noise diagnosis begins with a systematic appraisal of general noise location. Noises tend to telegraph from the actual source to the various chassis and body components. Air conditioning, alternator, power steering, and fan noises are often blamed on the transmission. One method of isolating transmission noise from accessory noises is to remove the drive belts from the accessories and drive the vehicle. If the noise disappears, it was in one of the accessories. Noises that vary with speed and road surface are most likely due to tires. Noises that vary with speed and direction are probably caused by wheel bearings. Coasting the vehicle in "neutral" with the engine shut off will eliminate the engine and transmission pump noises. Duplicating driving conditions with the vehicle on a hoist will eliminate tire noises. Running the vehicle with the drive shaft removed will eliminate drive line noise.

Shop Diagnosis

The last phase of troubleshooting involves a shop diagnosis. The shop diagnosis procedure may involve pressure testing and air checking. Pressure tests

BAND AND CLUTCH APPLICATION CHART				
LOW (Breakaway)	*LOW (Manual)*	*SECOND*	*DIRECT*	*REVERSE*
Rear Clutch	Rear Clutch	Rear Clutch	Rear Clutch	Front Clutch
Overrunning Clutch	Low and Reverse Band	Kickdown Band	Front Clutch	Low and Reverse Band

Figure 13-3

Band and Clutch Application Chart
(Chrysler Corp.)

Figure 13-4

Pressure Plug for Testing Operating Pressure

are made to determine if fluid pressures in the various circuits are sufficient for proper operation. The transmission manufacturer provides several plugs on the outside of the transmission case for pressure gage hook-up.

The technician connects the pressure gages to the transmission and either road tests the vehicle or tests it on a hoist to determine if pressures are within specifications. The manufacturer will provide a chart showing what each of the pressures

Figure 13-5

Pressure Chart
(Ford Motor Co.)

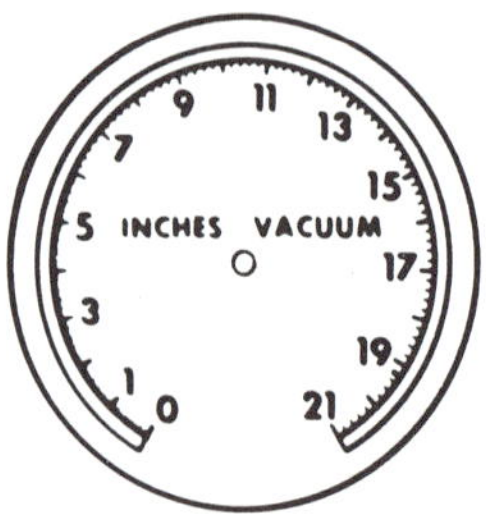

VACUUM GAGE

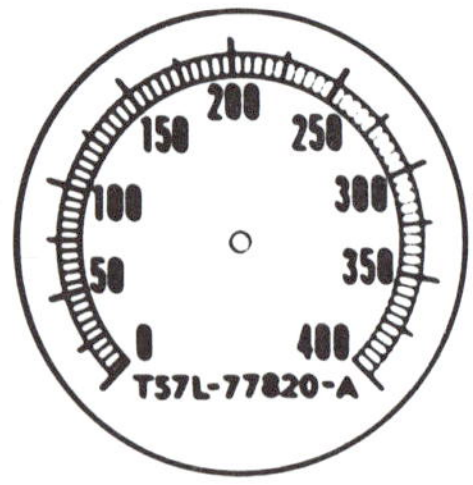

OIL PRESSURE GAGE

CONTROL PRESSURE SPECIFICATIONS

TRANSMISSION MODEL	THROTTLE POSITION	SELECTOR POSITION	VACUUM GAGE	PRESSURE GAGE
FX	IDLE	P,N,D,2,1	ABOVE 16	57-67
		R		57-84
	AS REQUIRED	D,2,1	10	67-106
	AS REQUIRED	D,2,1	BELOW 1	130-160
		R		179-213
MX EXCEPT PFA-E	IDLE	P,N,D,2,1	ABOVE 16	57-77
		R		71-106
	AS REQUIRED	D,2,1	10	85-122
	AS REQUIRED	D,2,1	BELOW 1	151-176
		R		201-213
MX PFA-E	IDLE	P,N,D,2,1	ABOVE 16	57-77
		R		64-105
	AS REQUIRED	D,2,1	10	85-122
	AS REQUIRED	D,2,1	BELOW 1	151-176
		R		196-213

should be. In this manner, the problem can be isolated to a particular circuit and a repair performed.

Air checking is a means of using air pressure to determine if the various units are in proper operating condition, and if their circuits are blocked or leaking. Air checking can be done with the transmission in the vehicle, usually with valve body removed, or with the transmission on the bench. Air checking is not only valuable as an aid to diagnosis, but also as a check on repairs. Air checking involves applying shop air to the apply passages for the various clutches and bands. The technician can listen for signs of blockage or leakage. He can hear the units apply. The manufacturer provides a chart showing where to apply the shop air.

Figure 13-6

Air Checking Chart

(Ford Motor Co.)

REVERSE-HIGH CLUTCH

Apply air pressure to the reverse-high clutch passage. A dull thud indicates that the reverse-high clutch piston has moved to the applied position. If no noise is heard, place the finger tips on the clutch drum and again apply air pressure to detect movement of the piston.

FORWARD CLUTCH

Apply air pressure to the transmission case forward clutch passage. A dull thud can be heard when the clutch piston is applied. If no noise is heard, place the finger tips on the input shell and again apply air pressure to the forward clutch passage. Movement of the piston can be felt as the clutch is applied.

INTERMEDIATE SERVO

Hold the air nozzle in the intermediate servo apply passage. Operation of the servo is indicated by a tightening of the intermediate band around the drum. Continue to apply air pressure to the intermediate servo apply passage, and introduce air pressure into the intermediate servo release passage. The intermediate servo should release the band against the apply pressure.

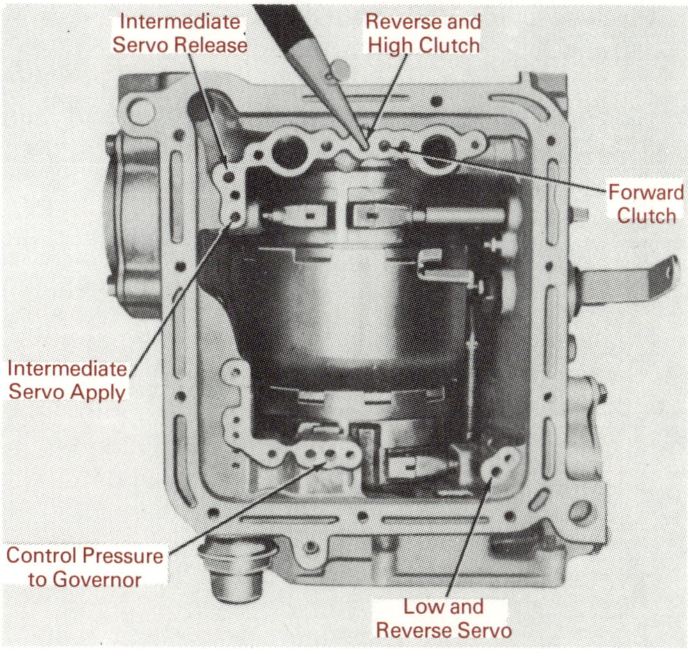

GOVERNOR

Apply air pressure to the governor passage and listen for a sharp clicking or whistling noise. The noise indicates governor valve movement.

LOW-REVERSE SERVO

Apply air pressure to the low-reverse apply passage. The low-reverse band should tighten around the drum if the servo is operating properly.

UNIT 13
Automatic Transmission Troubleshooting

True-False Questions (Indicate the correct choice)

1. When a technician encounters a transmission that is malfunctioning, the first thing he should do is road test the vehicle. T F

2. The automatic transmission specialist need not be concerned with engine operation. T F

3. Both too high and too low a fluid level will cause the transmission to slip. T F

4. If fluid smells burned, frictional material has been overheated. T F

5. If varnish is allowed to build up in a transmission for a long time, the transmission may stop operating. T F

6. Engine speed should change when the hose is pulled off the vacuum modulator. T F

7. A technician can isolate problems during a road test by using a band and clutch application chart. T F

8. Air-conditioning, alternator, power steering, and fan noises are often blamed on the transmission. T F

Multiple Choice Questions (Indicate the correct choice)

1. Troubleshooting the transmission involves:
   a. preliminary checks to eliminate the obvious
   b. road testing
   c. shop diagnosis
   d. all the above

2. A milky appearing fluid indicates:
   a. water in the fluid b. varnish in the fluid c. correct fluid appearance

3. A dip stick that is tacky and will not wipe clean indicates:
   a. wrong type of fluid b. varnish build-up c. correct fluid appearance

4. A transmission that creeps in neutral most likely has trouble with:
   a. throttle linkage c. manual control linkage
   b. vacuum modulator d. fluid level

5. Shop diagnosis involves:
   a. pressure testing   b. air checking   c. both *a* and *b*

UNIT 13
Automatic Transmission Troubleshooting

True-False Questions

1. This statement is false. With a malfunctioning transmission, the technician must first perform some preliminary checks to eliminate any obvious problems before road testing the vehicle.

2. This statement is false. The best operating transmission in the world will not make the vehicle perform if the engine cannot deliver the power.

3. This statement is true. If the fluid is low the pump will pick up air. If the fluid is too high, churning will cause foam. In both cases, the air results in slipping.

4. This statement is true. If fluid smells burned, the frictional material has been overheated. A complete overhaul will probably be necessary.

5. This statement is true. Eventually, the varnish formation will clog the filter or screen, pump pressure will drop, and the transmission may fail to operate completely.

6. This statement is true. The test of a vacuum modulator is to pull off the vacuum hose. The engine speed should change significantly.

7. This statement is true. A band and clutch application chart shows what is applied in each operating mode. The technician can then isolate the problem to a particular band or clutch.

8. This statement is true. These noises are often confused with transmission problems. One method of isolating transmission noise from accessory noises is to remove the accessory drive belts.

Multiple Choice Questions

1. The correct choice is *d.* Preliminary checking, road testing, and shop diagnosis are all performed in troubleshooting a transmission.

2. The correct choice is *a.* A milky appearing fluid indicates water is mixed with the fluid.

3. The correct choice is *b.* A dip stick that is tacky and will not wipe clean indicates a varnish build-up.

4. The correct choice is *c.* An indication that manual control linkage is out of adjustment is that the vehicle creeps in neutral.

5. The correct choice is *c.* The shop diagnosis procedure involves both pressure testing and air checking.

(Unit number refers to the unit in which the term is explained.)

Glossary

Accumulator Valve—Valve used to regulate fluid flow to and from servos. This valve directs flow to both sides of the servo pistons to achieve a smooth application and release of the band. (Unit 10)

Anti-rust—One of the properties of automatic transmission fluid. Air and water inside the transmission could cause rust. Fluids must prevent the formation of rust. (Unit 2)

ATF—Automatic Transmission Fluid. ATF is the standard abbreviation used by oil companies and transmission manufacturers to identify automatic transmission fluid. (Unit 2)

Band—Planetary holding unit. A band is wrapped around a planetary member. The planetary member is prevented from rotating when the band is applied or tightened around the member. (Unit 6)

Compatibility—One of the properties of automatic transmission fluid. Fluids for use in automatic transmissions must be compatible with the metallic and nonmetallic parts used in the transmission. (Unit 2)

Crescent Pump—Pump using a crescent-shaped seal to separate inlet from outlet. The crescent or IX pump is the most popular pump used in automatic transmissions. It consists of an inner gear keyed to the drive mechanism, a larger external gear, and a crescent seal installed in a close fitting housing. (Unit 8)

Detent Valve—Valve used to control forced downshifts for acceleration. The detent valve is connected to the accelerator pedal through mechanical or electrical linkage. (Unit 10)

Dexron—A type of automatic transmission fluid which has passed a given qualification requirement. Dexron is a fluid formulated to General Motors' specifications for their late-model transmissions. (Unit 2)

Downshift Valve—Same as **Detent Valve.**

Flow—Movement of fluid in a hydraulic system. Fluid flows in an automatic transmission by creating a difference in pressure between two points. (Unit 7)

Flow Rating—Number of gallons per minute a pump can deliver. Pumps for automatic transmissions must have a high flow rating to operate controls as well as to cool and lubricate. (Unit 8)

Fluid Coupling—A hydraulic drive. Early automatic transmissions used a fluid coupling to connect and disconnect the engine from the transmission. (Unit 3)

Foam Resistance—One of the properties of automatic transmission fluid. Automatic transmission fluid must resist foaming because foam would interfere with the fluid's ability to lubricate and transmit pressure. (Unit 2)

Front Pump—Pump located at the front of the automatic transmission. The front pump is driven by the engine through two dogs on the torque converter housing. The front pump supplies fluid whenever the engine is running. (Unit 8)

Gear Pump—Pump using two gears in mesh to carry fluid from inlet to outlet. A gear pump consists of two gears with external teeth meshed together in a housing. These pumps have become less popular because of their fluctuating delivery. (Unit 8)

Governor Valve or Governor Assembly—Device used to sense vehicle speed. The governor valve is attached to the transmission output shaft. Since the output shaft is connected to the drive shaft, the governor is able to sense vehicle speed for automatic upshifts and downshifts. (Unit 10)

Hydraulic Controls—One of the three main sections of an automatic transmission. The hydraulic control section is the brain of the automatic transmission. It senses and selects the correct degree of torque multiplication for every driving requirement. (Unit 1)

IX Pump—Same as **Crescent Pump.**

Lubricity—One of the properties of automatic transmission fluid. Lubricity refers to how well a fluid works to prevent wear between closely fitted working parts. Direct metal-to-metal contact must be avoided to prevent wear. (Unit 2)

Manual Control Valve—Valve used to manually select the operating mode of the transmission. The manual control valve is moved by linkage connected to the shifting quadrant inside the vehicle. The driver selects drive, low, park, reverse, etc., by moving the quadrant lever into position. (Unit 10)

Multiple Disc Clutch—Clutch using a number of discs to lock and unlock a planetary member. Multiple disc clutches use hydraulic pressure applied to a set of clutch discs to hold a planetary member. Their primary advantage is a great deal of frictional area in a small package. (Unit 6)

Non-positive Displacement Pump—A type of pump that has an inlet and outlet connected hydraulically. If a pump's inlet and outlet are connected so that fluid can recirculate in the pump, it is called non-positive displacement. These pumps are not used in automatic transmissions. (Unit 8)

Non-return Valve—Valve used to regulate flow. A non-return valve is a small steel ball located in a fluid passageway. Fluid in the passageway is allowed to move in one direction but prevented by the ball from going the other way. (Unit 9)

One-Way Clutch—A device which permits rotation in one direction only. The one-way clutch is composed of a set of rollers which roll on an inner and outer race. The rollers permit rotation in one direction, but jam the races and prevent rotation in the other direction. (Unit 6)

Overrunning Clutch—Same as **One-Way Clutch.**

Oxidation Resistance—One of the properties of automatic transmission fluid. Oxidation results when fluid combines chemically with oxygen. Oxidation must be prevented because it causes sludge. (Unit 2)

Planetary Gearbox—One of the three main sections of an automatic transmission. The planetary gearbox contains the planetary gears which provide a means of varying the gear ratio between the engine and the rear wheels. (Unit 1)

Planetary System (Simple)—A planetary system using a sun gear, a set of planets, a carrier, and a ring gear. Simple planetary gear sets have been used on many automatic transmissions. The unit allows gear ratio changes without taking any gears out of mesh. (Unit 5)

Planetary System (Complex)—A planetary system combining two or more simple planetary sets into one unit. A complex planetary system works like a simple planetary unit, but it is more flexible and occupies less space. (Unit 5)

Planet Carrier—Component of a planetary gear set. The carrier supports or carries the planets. (Unit 5)

Planets—Components of a planetary gear set. The planets are mounted in the carrier and revolve around the sun gear. (Unit 5)

Positive Displacement Pump—A pump that has its inlet sealed from its outlet. All automatic transmissions use fluid pumps of the positive displacement type. (Unit 8)

Pressure—The operating force of a hydraulic system. Pressure is the force used in an automatic transmission to move valves and lock planetary members. Pressure in a closed system is calculated by dividing the force applied by the area of the apply piston. (Unit 7)

Pressure Rating—The amount of pressure a pump can withstand without damage.

Pumps for automatic transmissions must have a high pressure rating because of the high pressures required to operate the transmission. (Unit 8)

Pressure Regulator Valve—Valve used to maintain a specific amount of pressure in a valve body. All the fluid entering the valve body must pass through the pressure regulator valve. This valve acts to bleed off pressure in excess of that required for valve body operation. (Unit 10)

Rear Pump—Pump located at the rear of an automatic transmission. The rear pump, if one is used, is driven by the output shaft. This pump can supply fluid whenever the rear wheels revolve. (Unit 8)

Relief Valve—Valve used to protect against excessive pressure. Valve body circuits are often protected against excessive pressure which might result if the normal pressure regulator should malfunction. (Unit 10)

Ring Gear—Component of a planetary gear set. The ring gear has internal teeth which mesh with the planets. (Unit 5)

Roller Clutch—Same as **One-Way Clutch.**

Scheduling Valve—Valve used to prevent two circuits from being applied at the same time. If two circuits were applied at the same time the transmission might slip. A scheduling valve prevents this by delaying the application of a circuit until the previous circuit can release. (Unit 10)

Servo—Piston assembly used to apply a band. Hydraulic pressure is directed at the servo piston which moves to squeeze the band around a planetary member. (Unit 6)

Shift Valve—Valve that directs an automatic upshift or downshift. The shift valve responds to engine load and vehicle speed to cause an upshift or downshift. (Unit 10)

Spool Valve—A cylindrically shaped valve with two or more valleys between the lands. Spool valves are used to direct flow. Fluid passages are opened or closed depending on the valve land positions. (Unit 9)

Sprag Clutch—Same as **One-Way Clutch.**

Suffix A—Automatic transmission fluid specification. Manufacturers developed the Suffix A specification in the 1950s when deficiencies were found in oxidation resistance in some Type A fluids. (Unit 2)

Sun Gear—Component of a planetary gear set. The sun gear gets its name from the central position it occupies in a planetary gear set. (Unit 5)

Throttle Valve—Valve used to sense how hard the engine is working. The throttle valve is connected to the accelerator pedal. This valve is able to sense how hard the engine is working, and sends a signal to the shift valve to regulate the shift point to help the engine. (Unit 10)

Timing Valve—Same as **Scheduling Valve.**

Torque—Turning effort applied to something to make it turn or try to make it turn. Torque is used in an automobile to rotate the rear wheels. The torque is developed by the engine and multiplied by the transmission. (Unit 4)

Torque Conversion—Torque multiplication using a hydraulic drive called a torque converter. A torque converter is able to multiply torque because stationary blades on the stator redirect fluid back toward the pump and help it turn. (Unit 4)

Torque Converter—Hydraulic drive designed to multiply torque. Most automatic transmissions use a torque converter. It serves the same function as a fluid coupling, with the additional advantage that it can increase torque. (Unit 4)

Torque Converter Pump—The driving or input member of a torque converter. The pump is connected to the engine. When the engine is running, fluid is thrown out of the pump to drive the turbine. (Unit 4)

Torque Converter Stator—The torque converter member that redirects fluid from the turbine back toward the pump. The stator acts as a reactionary element and is responsible for torque multiplication. (Unit 4)

Torque Converter Turbine—The driven or output member of a torque converter. The turbine is connected to the input shaft of the planetary gearbox. Fluid from the pump is directed at the turbine to make it turn. (Unit 4)

Type A—Automatic transmission fluid specification. Type A fluids were developed as an improvement over engine oils for use in automatic transmissions. (Unit 2)

Type F—Automatic transmission fluid specification. Type F fluid specifications were developed by Ford as an improvement over Type A, Suffix A fluids. (Unit 2)

Vacuum Modulator—Device used to sense engine load by monitoring engine vacuum. Most late-model transmissions have replaced the throttle valve with a vacuum modulator. The modulator does the same job as the throttle valve, but it is able to maintain a better match between the engine and transmission. (Unit 10)

Vane Pump—Pump using a series of spring-loaded vanes that sweep an eccentric cavity. Some automatic transmissions have used this pump, but it is expensive because of the large number of parts of which it is composed. (Unit 8)

Viscosity—The thickness or thinness of a fluid. Automatic transmission fluids must have a viscosity which allows them to flow freely in cold conditions and have sufficient body during high temperature operation. (Unit 2)

Viscosity Index—The amount that a fluid changes in viscosity with changing temperature. A fluid that changes relatively little with a given temperature change is said to have a high viscosity index. (Unit 2)

Vortex—Path of fluid flow in a torque converter or fluid coupling. The vortex may be high, low, or zero depending on the relative speed between the pump and turbine. (Unit 3)

Index